Springer Series in
MATERIALS SCIENCE 61

Springer
Berlin
Heidelberg
New York
Hong Kong
London
Milan
Paris
Tokyo

Springer Series in
MATERIALS SCIENCE

The Springer Series in Materials Science covers the complete spectrum of materials physics, including fundamental principles, physical properties, materials theory and design. Recognizing the increasing importance of materials science in future device technologies, the book titles in this series reflect the state-of-the-art in understanding and controlling the structure and properties of all important classes of materials.

61 **Fatigue in Ferroelectric Ceramics and Related Issues**
By D.C. Lupascu

62 **Epitaxy**
Physical Foundation and Technical Implementation
By M.A. Herman, W. Richter, and H. Sitter

63 **Fundamentals of Ion Irradiation of Polymers**
By D. Fink

64 **Morphology Control of Materials and Nanoparticles**
Advanced Materials Processing and Characterization
Editors: Y. Waseda and A. Muramatsu

65 **Transport Processes in Ion Irradiated Polymers**
By D. Fink

66 **Multiphased Ceramic Materials**
Processing and Potential
Editors: W.-H. Tuan and J.-K. Guo

67 **Nondestructive Materials Characterization**
With Applications to Aerospace Materials
Editors: N.G.H. Meyendorf, P.B. Nagy, and S.I. Rokhlin

68 **Diffraction Analysis of the Microstructure of Materials**
Editors: E.J. Mittemeijer and P. Scardi

69 **Chemical-Mechanical Planarization of Semiconductor Materials**
Editor: M.R. Oliver

70 **Isotope Effect Applications in Solids**
By G.V. Plekhanov

71 **Dissipative Phenomena in Condensed Matter**
Some Applications
By S. Dattagupta and S. Puri

72 **Predictive Simulation of Semiconductor Processing**
Status and Challenges
Editors: J. Dabrowski and E.R. Weber

Series homepage – springer.de

Volumes 10–60 are listed at the end of the book.

Doru C. Lupascu

Fatigue in Ferroelectric Ceramics and Related Issues

With 80 Figures

Springer

Dr. Doru C. Lupascu
Technische Universität Darmstadt
Fachbereich Materialwissenschaften
Nichtmetallisch-Anorganische Werkstoffe
Petersenstrasse 23
64287 Darmstadt, Germany
E-mail: lupascu@ceramics.tu-darmstadt.de

Series Editors:

Professor Robert Hull
University of Virginia
Dept. of Materials Science and Engineering
Thornton Hall
Charlottesville, VA 22903-2442, USA

Professor R. M. Osgood, Jr.
Microelectronics Science Laboratory
Department of Electrical Engineering
Columbia University
Seeley W. Mudd Building
New York, NY 10027, USA

Professor Jürgen Parisi
Universität Oldenburg, Fachbereich Physik
Abt. Energie- und Halbleiterforschung
Carl-von-Ossietzky-Strasse 9–11
26129 Oldenburg, Germany

Professor Hans Warlimont
Institut für Festkörper-
und Werkstofforschung,
Helmholtzstrasse 20
01069 Dresden, Germany

ISSN 0933-033X

ISBN 3-540-40235-7 Springer-Verlag Berlin Heidelberg New York

Cataloging-in-Publication Data applied for.
Bibliographic information published by Die Deutsche Bibliothek Die Deutsche Bibliothek lists this publication in the Deutsche Nationalbibliografie; detailed bibliographic data is available in the Internet at http://dnb.ddb.de.

Springer-Verlag is a part of Springer Science+Business Media.

springeronline.com

Typesetting by the author
Cover concept: eStudio Calamar Steinen
Cover production: *design & production* GmbH, Heidelberg

Printed on acid-free paper 57/3141/tr 5 4 3 2 1 0

To Kathrin, my parents, and my brother

Preface

Ferroelectric devices are nowadays widely found in advanced technology as well as everyday household machinery. The broad range of physical properties in this class of materials has allowed for the evolution of devices ranging from piezoelectric components and optical devices to ferroelectric non-volatile memories in microelectronics. Examples extend from simple gas igniters, sonic and ultrasonic sensors and actuators to the more advanced ultrasonic motors, high precision positioning devices, optical switches and recently permanent memories for computer applications. Many more applications are anticipated. Particularly the fact that many parts are at the edge of becoming large numbered consumer products necessitates a much higher reliability of the devices. Two aspects are relevant with respect to this, an adequate device design to ensure a minimum loading of the ferroelectric material on the one hand, and optimized material properties for high reliability at maximum performance on the other. Unfortunately, ferroelectric materials are mostly brittle both as single crystals as well as polycrystalline material, and in the worst case even water soluble crystals. Particularly tensile mechanical loading will often lead to an immediate failure of the device. On longer terms, cyclic loading will deteriorate the performance of the material. Multiple loading will either lead to cyclic crack growth or the modification of the material properties at the mesoscopic and microscopic level. During the last fifty years of material development it has become clear that effects on a wide range of time and size scales down to the atomic level are relevant for the properties of ferroelectrics. In these same ranges of times and sizes the sources of fatigue have to be identified. The present text is an attempt to assign the fatigue phenomenon in bulk ferroelectric ceramics to its microscopic, mesoscopic, and macroscopic sources.

Darmstadt, October 2003 *Doru Constantin Lupascu*

Acknowledgement

The largest difficulty in writing a book resides in an appropriate appreciation of the work of all those who have contributed to its successful completion. I would like to express my gratitude to all my colleagues and friends who have provided insight, generosity, numerous work hours, and plenty of joy during the last years. This book is the outcome of all this collaborate effort.

First of all, I want to express my gratitude to Jürgen Rödel for his confidence in my abilities and the continuous intellectual, financial, and personal support which have entered this and many other projects throughout the last years.

Next, I am very grateful to Jürgen Nuffer, Cyril Verdier, and Markus Christman, who were directly involved in the fatigue studies and achieved many of the experimental results presented in this book in long laboratory hours.

Furthermore, I would like to express my thankfulness to Sergio Louis dos Santos e Lucato and Alain Kounga, who have successfully worked on so many aspects of fracture of ferroelectrics. Achim Neubrand has been my brilliant office mate.

The entire team of Nichtmetallisch-Anorganische Werkstoffe provided an ongoing pleasure in the daily work. Emil Aulbach, Markus Bürgers, Astrid Dietrich, Matthias Diemer, Jens Eichler, Susan Galal-Yousef, Roswitha Geier, Herbert Hebermehl, Daniel Hoffmann, Mark Hoffmann, Ralf Jedamzik, Johannes Kanters, Patricie Merkert, Thomas Ostrowski, Siegfried Skirl, Martin Stech, Ulrich Stiefel, Thomas Utschig, Ludwig Weiler, Rongjun Xie, André Zimmermann, Ruzhong Zuo, and many others have contributed to the fun in our group and the excellent working environment. Our visitors Didier Rescamp, Arnab Chattopadhyay, Thomas Karastamatis, Jaqueline Menchaca, William Oates, and many others brought their work and the international stimulus into our group. Furthermore I would like to welcome our new team members Nina Balke, Thorsten Schlegel, and Ilona Westram in their endeavors of ferroelectrics research.

I would then like to very much appreciate the exceptional collaboration with Christopher S. Lynch from the George Woodruff School of Mechanical Engineering at the Georgia Institute of Technology, Atlanta, Georgia. His and his family's hospitality during my several visits to Atlanta and the long lasting scientific conversations in the lab, in the office, and late at night made me feel welcome and respected wherever and whenever we met.

I thank Qiming Zhang from the Materials Research Laboratory at the Pennsylvania State University, who hosted me during my first footsteps into ferroelectricity and gave me the possibility to enjoy the spirit of the largest school in ferroelectrics in the world. I furthermore thank Hideaki Aburatani from the same school for sharing his knowledge on acoustic emission of ferroelectrics. And I thank the entire team at the MRL in PennState for much fun and the tale-gating in the snow.

I then thank Vladimir and Alevtina Shur from the Ural State University in Ekatherinburg in Russia for the friendly and fruitful collaboration on the discontinuous processes in ferroelectric single crystals and the excellent crystals they provided for our measurements.

Other collaborators from all over the world have contributed: Marianne Hammer, Axel Endriss, Alexandre Glazounov, Michael Hoffmann and Jan Reszat from the Universität Karlsruhe, Ute Bahr, Hans-Achim Bahr, Herbert Balke, Bak Pham from the Technische Universität Dresden, Marc Kamlah, Theo Fett, and Dietrich Munz from the Forschungszentrum Karlsruhe, Ute Rabe and Walter Arnold from the Universität Saarbrücken, Hans-Joachim Kleebe from the Universität Bayreuth, Michael Schröder from the Institut für Physkalische Chemie, TU Darmstadt, and Markus Kreuzer from the Institut für Angewandte Physik, TU Darmstadt. I would also like to acknowledge our new and already so fruitful collaboration with Jim Scott and his team from Cambridge.

I am then very much indebted to the Institut für Materialwissenschaften for the excellent working environment concerning equipment, personnel, and the flexibility in work procedures in many respects. No experimental work is successful without a good workshop. Jochen Korzer and his team have furnished an outstanding quality of equipment for all our experimental ideas.

All members of the Schwerpunktprogramm Multifunktionswerkstoffe of the Deutsche Forschungsgemeinschaft have provided useful discussions and happy evenings.

The financial support by the Deutsche Forschungsgemeinschaft is greatly acknowledged.

My parents are at the source of it all. They raised me lovingly and initiated an interest in science and technology in an early stage of my life. My brother has been a constant joy throughout.

Last, but not least, I would like to express my love to my wife Kathrin, who has been so patient in the course of manuscript writing and beyond.

Contents

Symbols and Abbreviations XV

1 Introduction 1
1.1 Document Structure 1
1.2 Fatigue 1
1.3 Historical 2
1.4 Ferroelectricity and Piezoelectricity 4
1.5 The Lead-Zirconate-Titanate Crystal System 8
1.5.1 Dopants 8
1.5.2 PIC 151 10
1.5.3 Lanthanum Doped PZT 11
1.6 Defects in PZT 14
1.6.1 Vacancies and Interstitials 14
1.6.2 Thermal Equilibrium 14
1.6.3 The Quenched State 16
1.7 Macroscopic Fatigue 18
1.8 Point Defects and Fatigue 20
1.9 Domain Pinning, Aging, and Imprint 23
1.10 Electrodes and Thin Films 27
1.11 Crystal- and Microstructure 32
1.12 Temperature Dependence of Fatigue 33
1.13 Switching, Relaxation, and Rate Dependencies 34
1.14 Microcracking 35
1.15 Models 36
1.16 Fatigue-Free Systems 39

2 Macroscopic Phenomenology 41
2.1 Fatigue and Measurement Procedures 41
2.2 Polarization and Strain Loss 43
2.3 Asymmetry and Offset-Polarization 46
2.3.1 Strain Asymmetry 46
2.3.2 Obstacles to 90° Domain Switching 48
2.4 Anisotropy 51
2.5 Leakage Current, Sample Coloring, and Relaxation 55

2.6 Unipolar Fatigue ... 57
2.7 Mixed Loading Fatigue ... 58

3 **Agglomeration and Microstructural Effects** ... 63
3.1 Agglomerates ... 63
3.2 Grain Boundaries ... 63
3.3 Agglomeration and Crystal Structure ... 70
3.4 Domain Structure ... 72
3.5 Unit Cell Volume ... 73
3.6 The Oxygen Balance ... 74
3.7 Microcracking ... 75
3.7.1 Edge Effects ... 76
3.7.2 Bulk Microcracking ... 77
3.8 Macrocracking, Delamination Fracture ... 78

4 **Acoustic Emission and Barkhausen Pulses** ... 81
4.1 Acoustic Emission Technique ... 81
4.1.1 Principle ... 81
4.1.2 Instrumentation ... 82
4.2 Polycrystalline Lead-Zirconate-Titanate ... 84
4.2.1 Acoustic Emission Sources ... 84
4.2.2 Different Crystal Structures of PZT ... 85
4.2.3 Fatigue Induced Discontinuities ... 96
4.2.4 Acoustic Emissions under Uniaxial Stress ... 97
4.3 Single Crystal Ferroelectrics ... 101
4.3.1 The Uniaxial Ferroelectric Gadolinium Molybdate ... 101
4.3.2 The Perovskite Type Barium Titanate ... 107

5 **Models and Mechanisms** ... 111
5.1 Fatigue Models for a Polycrystalline Ferroelectric ... 111
5.2 Band Structure ... 114
5.3 Point Defects and Dipoles ... 116
5.3.1 Concentration ... 116
5.3.2 Defects and Microdomains ... 117
5.3.3 Localized Electron States ... 118
5.3.4 Defect Dipoles ... 121
5.4 Ion and Electron Motion ... 121
5.4.1 Diffusion Within a Domain ... 121
5.4.2 Anisotropy and Directionality of Diffusion ... 122
5.4.3 Drift and Convection ... 128
5.4.4 Directionality of Electron Motion ... 129
5.4.5 Average Conductivities ... 129
5.5 Screening, Space Charges, and Domain Freezing ... 130
5.6 Electrodes ... 134
5.7 Grain Boundaries ... 136

5.8 Agglomeration 137
5.8.1 Point Defect Sinks 137
5.8.2 Dislocation Loop Formation at Low p_{O_2} 139
5.8.3 Induced Crystallographic Phase Change 144
5.8.4 Iterative Models in General 146
5.8.5 Iterative Model of Bulk Agglomeration 151
5.8.6 Rapid Fatigue due to Polarons 159
5.9 Crystallite Orientation and Anisotropies 161
5.10 Relaxation Times 164
5.11 Unclear Effects 164

6 Recent Developments 167
6.1 Material Modifications 167
6.1.1 Excess and Deficient Lead Oxide 167
6.1.2 Doping 168
6.1.3 Secondary Bulk Phases 168
6.1.4 Reduction of PZT, SBT, and BiT 169
6.2 Grain Boundary and Grain Size 170
6.2.1 Grain Boundary 170
6.2.2 Grain Size 170
6.3 Several Layers of Different Composition 171
6.3.1 Lead Excess Layers Near the Electrode 171
6.3.2 Multiple Ferroelectric/Antiferroelectric Layers, Buffer Layers 171
6.4 New or Modified Modelling Approaches 172
6.4.1 Relation to aging 172
6.4.2 Ising and Preisach Approach 172
6.4.3 Phase Transition Picture 173
6.4.4 Supplemental Arguments to Previous Chapters 173
6.4.5 Curved Domain Walls 176
6.4.6 Other Approaches 176
6.5 Leakage Current 176
6.5.1 Ionic Currents 176
6.5.2 Electronic States 177
6.5.3 Breakdown 178
6.6 Mechanical Effects 178
6.6.1 Mechanical Fatigue 178
6.6.2 Microcracking 179
6.7 Anisotropy of Fatigue 180
6.8 Time and Relaxation Effects 180
6.9 Electrode Geometry 182
6.10 Layered Perovskite Ferroelectrics 182
6.10.1 Strontium Bismuth Tantalate 182
6.10.2 Bismuth Titanate 184
6.11 Uniaxial Ferroelectrics 185

7 Summary 187

A Solutions to Integrals 193

References 195

Index 221

Symbols and Abbreviations

a.c.	Alternating current
AE	Acoustic emissions
AFM	Atomic force microscopy
BHP	Barkhausen pulses
BiT	Bismuth titanate ($Bi_4Ti_3O_12$)
BLT	Lanthanum substituted bismuth titanate ($Bi_{3.25}La_{0.75}Ti_3O_{12}$)
BNdT	Bismuth titanate ($Bi_4Ti_3O_{12}$)
$Bi_{4-x}Nd_xTi_3O_{12}$ (BNdT) BTO	Bismuth titanate ($Bi_4Ti_3O_{12}$)
d.c.	Direct current
EELS	Electron energy loss spectroscopy
EPR	Electron paramagnetic resonance
ESR	Electron spin resonance (=EPR)
FEM	Finite element modeling
GMO	Gadolinium molybdate ($Gd_2(MoO_4)_3$)
HRTM	High resolution transmission electron microscopy
LSCO	Lanthanum strontium cobaldate $(La, Sr)CoO_3$
PBN	Lead-barium-niobate ($Pb_xBa_{1-x}Nb_2O_3$)
PLZT	Lanthanum doped lead-zirconate-titanate ($La_yPb_{1-y}(Zr_xTi_{1-x})O_3$)
PMN	Lead magnesium niobate ($Pb(Mg_{1/3}Nb_{2/3})O_3$)
p_{O_2}	Oxygen partial pressure
PT	Lead-titanate ($PbTiO_3$)
PZN	Lead-zink-niobate ($Pb(Zn_{1/3}Nb_{2/3})O_3$)
PZT	Lead-zirconate-titanate ($Pb(Zr_xTi_{1-x})O_3$)
RTA	Rapid thermal annealing
SBN	Strontium barium niobate ($Sr_{1-x}Ba_xNb_2O_6$)
SBT	Strontium bismuth tantalate ($SrBi_2Ta_2O_9$)
SEM	Scanning electron microscopy
$\Sigma3$	Crystallographic coincidence lattice twin of period 3

SIMS Scanning ion mass spectrometry
TEM Transmission electron microscope
UV Ultraviolet
$V_O^{\bullet\bullet}$ Twice positively charged oxygen vacancy
XRD X-ray diffraction
XPS X-ray excited photoelectron spectroscopy
YBCO Yttrium barium copper oxide ($YBa_2Cu_3O_7$)
$\otimes$ Dyadic or outer product (Malvern, 1969)
$^{\bullet}$, $'$, $^{\times}$ Positively, negatively, and un-charged ion with respect to lattice position in an ionic crystal according to Kröger and Vink (1956)

1 Introduction

1.1 Document Structure

This book is divided into the following parts. The introduction (Chap. 1) defines the effects, gives a brief historical introduction, presents the materials investigated, and then outlines the present knowledge on the fatigue effect in ferroelectrics. The following chapters (Chaps. 2 and 3) fairly densely display the experimental results found in ferroelectric ceramics and only briefly discuss them in the context of previously known results. A larger section is devoted to the acoustic emission technique (Chap. 4). Chapter 5 then extensively discusses the possible explanations of the known and new experimental results in the context of existing and new fatigue models. Chapter 6 summarizes the recent developments in the field that have been published since the first version of this book was written. The contents are briefly summarized in Chap. 7.

For the reader less familiar with fatigue in ferroelectrics it may be advisable to first read the summary and introduction to see where the journey is going, and then return to the experimental chapters.

1.2 Fatigue

The Challenge

The challenge in studying fatigue stems from the high potential technological relevance of ferroelectric ceramics, and from the very controversial discussion of the underlying mechanisms during the last twenty years. Many aspects of the fatigue phenomenon have been revealed, but many facets are still only vaguely explicable by the mechanisms discussed so far. The vital relevance of assigning the most fatal fatigue mechanism, the loss of switchable polarization, to all its microscopic sources remains a frontier to be conquered.

Definition

Fatigue is defined as the, mostly unwanted, property changes of a material under cyclic external loading. Classically, this concerns fatigue of metals. Two forms of fatigue are generally considered in metals, fatigue due to embrittlement without preexisting cracks and fatigue crack growth.

In electroceramics the term fatigue has been used to denote effects that are brought about by cyclic electric loading or in few cases thermal or mechanical cycling. All of these may lead to the reduction in material performance like increasing conductivity for insulators and capacitor materials, less charge capacity in rechargeable batteries or capacitors, altered current-voltage characteristics in nonlinear resistances, or eVen to dielectric breakdown in any of these devices. In ferroelectrics the fatigue either denotes the reduction of switchable polarization or the reduction of the piezoelectric coefficient. Both quantities are closely related (see Sect. 1.4) and a common fatigue source possibly exists.

Most known ferroelectrics are also ferroelastic. They experience strain changes due to the piezoelectric effect and due to the switching of internal domains. Both the piezoelectric effect as well as the ferroelastic switching couple electric and mechanical properties. Fatigue may thus be induced by electric or mechanical cycling. As a consequence mechanical and electrical properties may degrade differently or concurrently just depending on the particular microscopic mechanism. In any case, the electromechanical coupling must be considered.

Most fatigue studies in ferroelectrics have been concerned with ferroelectric ceramics or thin films, but many of the effects equally well occur in single crystals. The latter are usually just too valuable to be deteriorated in experiment.

Fatigue has to be strictly distinguished from aging. While fatigue requires the repetitive application of external loads, aging is a pure time effect under constant external parameters. Aging occurs solely due to thermally excited processes, while fatigue may also be due to the external driving forces, mechanical embrittlement, micro- and macrocracking.

The experiments shown in this work are concerned with electric cycling of ferroelectric ceramics. One experiment also considers a mixed loading. Experiments on single crystals are only performed in the context of the acoustic emission studies. Measurements on single crystal lead zirconate titanate are simply not possible as such crystals are non existent.

1.3 Historical

A summary of the historical development of research in ferroelectrics can be found in the book by Lines and Glass (1977), which is so well treated there that any further attempt to do so will necessarily fail. In the following only the works concerning fatigue are mentioned.

Aging was discovered early in the study of perovskite ferroelectrics. McQuarrie (1953) was the first to observe the deformation of the polarization hysteresis to a propeller-shaped loop in ceramic $BaTiO_3$ after several days of aging. This process is also sometimes called shelf-aging to distinguish it from fatigue-aging, which was used as a term in the early literature. The latter term should nowadays be strictly avoided in order to avoid confusion.

The first work to actually consider the influence of alternating fields on material behavior, is reported by Merz and Anderson (1955), who studied the fatigue effect in single crystal $BaTiO_3$. The influence of point defects on the fatigue mechanism was then anticipated by Anderson et al. (1955), who already discovered that $BaTiO_3$ would loose its square hysteresis loop when cycled in atmospheres which are now known to permit the incorporation of oxygen vacancies. In dry air and pure O_2 the cycling did not affect the hysteresis. Taylor (1967) then did the first fatigue studies on different compositions of niobium-doped $Pb(Zr, Sn, Ti)O_3$ and later on ferroelectric optical switches (Taylor et al., 1972). Stewart and Cosentino (1970) studied soft doped PZT (La and Bi) in optically transparent birefringent PLZT (2/65/35) and $Pb_{0.99}(Zr_{0.65}Ti_{0.35})_{0.98}O_3$ in more detail. They found that irrespective of electrodes, curve shape of the applied bipolar electric field, and ambient conditions the material fatigued under bipolar cycling at 60 Hz. They observed the reduction of remnant polarization, the increase in coercive field and a deterioration of the optical properties. They reported the loss in squareness of the hysteresis loop as well as the significant changes in the switching current characteristics loosing its peak around E_c. Also the possibility of partial recovery of the material behavior by thermal treatment at high temperatures or re-cycling at slightly elevated temperatures (150°C) for about 5% of the initial fatiguing time was reported. They already ruled out microcracking as the sole reason of fatigue. Fraser and Maldonado (1970) then reported a significant effect of electrodes on the fatigue behavior, which has become common agreement in literature now (see Sect. 1.10). Serious micromechanical effects were discovered to be at the origin of fatigue in La and Mn-doped $PbTiO_3$ (Carl, 1975) due to the high strains in this material. Salaneck (1972) showed that in 8/65/35 PLZT the fatigue of the polarization P and of the birefringence Δn are linearly related. EVen though microcracking was observed, he found that simple microcracking scenarios are not entirely appropriate to account for the loss of birefringence. He reported a logarithmic loss of polarization after about $5 \cdot 10^5$ cycles and an increase of the coercive field by about 6% after 10^7 cycles (Taylor et al., 1972). The subject then was addressed in a few works in the 80's, but increased interest in the topic only developed in the 90's in the context of ferroelectric memories and ferroelectric multilayer actuators. It was not until then, that the microscopic sources of the fatigue mechanism were seriously discussed.

1.4 Ferroelectricity and Piezoelectricity

Ferroelectric Hysteresis

Ferroelectricity is an effect occurring in certain insulating or semiconducting dielectrics. A ferroelectric develops a remnant change in polarization upon application of an external field (see e.g. the first row in Fig. 1.6). In almost all ferroelectric systems known, ferroelectricity develops below a certain temperature, the Curie-point, starting from a paraelectric high temperature phase, called the prototype phase. Microscopically the phase transition is achieved by the motion of ions in the unit cell between different sTable crystallographic positions. Some of these positions become energetically more favorable just by the application of an external electric field. Some of these transitions are displacive, others rotational. The switching between the different localizations requires a certain threshold termed the coercive field. Beneath this field a single unit cell will not switch. This is true in ideal crystals, where the switching is actually a very abrupt eVent (comparable to the rhombohedral hysteresis loop in Fig. 1.6). The most common description of this material behavior without any reference to microscopic mechanisms, i.e. the electron orbitals of the ions involved, is the expression of the electric Gibbs's free energy $G_1 = G + DE$ of the system by a polynomial expression in polarization

$$G_1(T,P) = G_1(T) + \frac{1}{2}\beta(T)P^2 + \frac{1}{4}\xi P^4 + \frac{1}{6}\zeta P^6 + \cdots \qquad (1.1)$$

with the electric field E, the dielectric displacement D, and the polarization P, shown in the uniaxial case for simplicity. It is the Landau-Devonshire theory of ferroelectrics. $\boldsymbol{E}$, $\mathbf{D}$, and $\mathbf{P}$ are related via

$$\mathbf{D} = \epsilon_o \mathbf{E} + \mathbf{P} \ , \qquad (1.2)$$

where $\mathbf{P}$ itself is a function of the electric field. The derivatives of G yield all the other thermodynamic quantities to a fascinating degree of precision, if the appropriate coefficients are chosen (Devonshire, 1954; Mitsui et al., 1976; Lines and Glass, 1977; Xu, 1991).

The electric field becomes a polynomial in P containing only odd exponents:

$$E(P) = \beta(T)P + \xi P^3 + \zeta P^5 + \cdots \ . \qquad (1.3)$$

As the inverse function, $P(E)$, is the physically observed one, the minima and maxima of the polynomial become instability points resulting in the development of the hysteresis. The instability points are identical to the coercive fields, at least for a single unit cell. The polarization at $E = 0$ is the remnant polarization $P_r = P(E = 0)$. In the case of a single unit cell it is equal to the spontaneous polarization $P_s = P_r$. The spontaneous polarization is the result of a crystallographic phase transition from the centered position to an off-center position of a charge-carrying ion generating a local dipole in the unit cell, which is equivalent to an effective polarization. As the word instability point already suggests, the interior of the hysteresis is thermodynamically not very well determined. Nature thus developed many possibilities to access the interior of the hysteresis loop, particularly by the formation of domains. While in an ideal case, the macroscopic hysteresis is a one-to-one image of the unit cell hysteresis, in all practical cases the macroscopic hysteresis differs substantially from the microscopic one. Nevertheless, the formal description is applicable far beyond the unit cell case and commonly used for the macroscopic properties also. These basic facts are recited here, because the fatigue phenomenon is the massive influence of global or local boundary conditions shifting the material behavior back and forth within this maximum thermodynamically achievable hysteresis loop. It determines the boundaries of the thermodynamic and kinetic playground for fatigue.

Strain, Electrostriction, and Piezoelectricity

An extension of (1.1) can also account for the strains developed in the material via the introduction of the electrostrictive coefficients. In order to do so the next extensive variable strain S_{ij} (tensor or rank 2) has to be incorporated into the thermodynamic potential. This is achieved in the Helmholtz free energy $A(S,P) = G_1 + T_{ij}S_{ij}$

$$A(S,P) = \frac{1}{2}\beta_{ij}^{S} P_i P_j + \frac{1}{2}c_{ijkl}^{P} S_{ij} S_{kl} + \frac{1}{2}q_{ijkl} S_{ij} P_k P_l \qquad (1.4)$$

only showing terms to the order of 2. The stresses T_{ij} are given directly through the derivatives of A

$$T_{ij} = \frac{\partial A(S,P)}{\partial S_{kl}} = c_{ijkl}^{P} S_{kl} + q_{ijkl} P_k P_l \qquad (1.5)$$

retaining the indices $i, j, k, l = 1, 2, 3$ for the three directions in space. The c_{ijkl}^{P} are the elastic constants at constant P and the q_{ijkl} some electrostrictive coefficients. Again the inverse has to be used, thus

$$S_{ij} = s_{ijkl}^{P} T_{kl} + Q_{ijkl} P_k P_l \qquad (1.6)$$

with a different set of electrostrictive coefficients Q_{ijkl}. This equation is the heart of all strain coupling involved with ferroelectrics. In actuator applications the change of strain with respect to moderate changes in electric field is used and the property needed is the piezoelectric coefficient

$$d_{kij} = \frac{dS_{ij}}{dE_k} \,. \qquad (1.7)$$

For only small changes in electric field the change in polarization is given by its derivative, thus

$$\frac{dP_i}{dE_j} = \epsilon_o \epsilon_{ij;r}(P(E), T) , \tag{1.8}$$

where $\epsilon_{ij;r}(P(E), T)$ itself becomes a function of the polarization. From this and (1.6) the piezoelectric coefficient is given by:

$$d_{kij}(P(E), T) = 2Q_{ijkl}\, \epsilon_o \epsilon_r(P(E), T) \cdot P_l(E, T, \text{prehistory}) , \tag{1.9}$$

which is the material property exploited in actuator applications. While the electrostrictive coefficient is truly a material property only determined by the crystal structure and the electronic orbitals of the lattice atoms, the momentary polarization is a function of all the other variables, including all the microstructural effects, the prehistory, point defects and external boundary conditions. It is massively subject to fatigue. To some extent also the dielectric, particularly the macroscopic dielectric coefficient, will change during fatigue, all of which entering the actuator performance (Xu, 1991; Uchino, 1997).

Depolarizing Field and Domains

During the crystallographic transformation from the high temperature prototype phase to the ferroelectric phase the spontaneous polarization increases in one particular direction in the crystal. As long as no external free charge carriers are available, i.e. $D = 0$, this generates the so called depolarizing field through the expression

$$E_{depol} = -\frac{P_s}{\epsilon_o \epsilon_r} , \tag{1.10}$$

which stems directly from (1.2) and the linear approximation of the polarization dependence on field $P(E) = P_s + \epsilon_o \chi E$ and $\epsilon_r = 1 + \chi$. The product $dW = \int E \cdot dP$ is the volume density of energy stored in the electric field. The total energy of the system becomes proportional to the volume V and $P_s^2/(2\epsilon_o \epsilon_r)$, because no fraction of the electric field is compensated. Beyond a certain volume this becomes unbearable for the material and the domain system develops. Through the formation of domains the electric fields only reach from endpoint to endpoint of the respective domains, thus minimizing the volume in which large electric fields prevail. The size of the domains stays finite, because the domain walls develop simultaneously, bearing an energy area density of their own. A certain domain width minimizes the total energy between the field and domain wall energy.

Screening

Another possibility for the material to reduce its energy is the incorporation of free charges at the endpoints of the domains. In a monodomain material electroded at the two faces at tip and tail of the domain, electrons yield

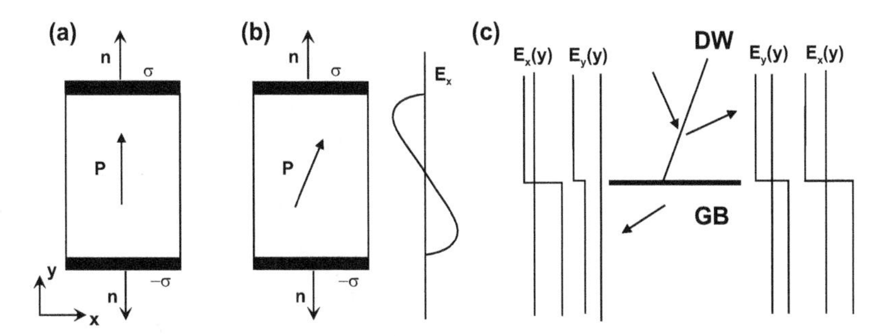

Fig. 1.1. Fields due to the depolarizing field in the microstructure. (**a**) Ideally compensated monocrystal, (**b**) field components despite external charges, and (**c**) complex fields arising at a grain boundary

the area charge density σ on the electrodes, which can fully compensate the depolarizing field via

$$\mathbf{D} = \sigma \cdot \vec{n} \tag{1.11}$$

with $\vec{n}$ being the normal to the electrode plane, if the electrodes are shorted (Fig. 1.1a). The system energy is zero. Equation (1.11) is the integrated Maxwell's law, $\mathrm{div}\mathbf{D} = \rho_{\mathrm{free}}$, for the boundary conditions of Fig. 1.1a. Once the orientation of the polarization in the crystallite is no longer perfectly orthogonal to the electrodes, electric field components will arise in planes parallel to the electrode plane (Fig. 1.1b). At the electrode itself, these fields are shorted by currents trough the electrode. The transverse fields in the bulk at locations remote from the electrode already become complicated solutions to Maxwell's law. A comparison of such theoretical derivations and actual charge states determined from the mobility of domains under an electron beam of a transmission electron microscope in $BaTiO_3$ were recently given by Krishnan et al. (2000b) (see the discussion of the interaction of domains walls with extended charges in Sect. 5.8.5).

In the case of a grain boundary, the field situation may become very complex. Figure 1.1c shows a possible configuration of domains around a grain boundary along with the electric fields. The grain boundary is considered far away from any conducting surfaces and all fields fully develop. It is eVident that the incorporation of free positive charge carriers reduces the total energy particularly at the right side of the image. In addition, due to charge states which are particular to grain boundaries, the situation is further complicated (see Sect. 5.7).

Many different words have been chosen to describe the influence of extra charge carriers in the microstructure, screening field, bias filed, local bias field, internal field, just to name a few. To my opinion, the only difference between these terms is the length scale at which the free charges modify the local fields. They are all simply expressing the fact that free charge carriers are more or less located within the insulating or semiconducting ferroelectric and, via their charge, modify the response of the domain system to external fields.

The differences in the resulting material behavior moreover stem from the condition, whether these extra charges are capable of blocking the domains from moving, or whether they just modify the external field by some amount. This point will be discussed in detail in Sect. 5.5. It is closely related to some fatigue scenarios.

1.5 The Lead-Zirconate-Titanate Crystal System

The binary phase diagram for lead titanate and lead zirconate is shown in Fig. 1.2. The technologically most relevant feature of this phase diagram is the morphotropic phase boundary between the rhombohedral structure on the zirconium rich side of the diagram and the tetragonal phase on the titanium side. Similar to the phase boundary to the cubic prototype phase above the Curie point most material parameters show extreme values near the morphotropic phase boundary (Jaffe et al., 1971). This particularly concerns the dielectric constant and the piezoelectric coefficient. Both effects are basically due to the high polarizability of the material at the morphotropic phase boundary also yielding a high remnant polarization. The latter is directly exploited in memory applications. Indirectly it also determines the piezoelectric coefficient via (1.9). The choice of material compositions near the morphotropic phase boundary is eVident. Further details can be found e.g. in Jaffe et al. (1971) and Xu (1991).

1.5.1 Dopants

To further stabilize the behavior of the material at this morphotropic phase boundary dopants are added. Donor dopants will render the material "soft" and acceptors "hard". Soft dopants reduce the concentration of oxygen vacancies $V_O^{\bullet\bullet}$ and increase the concentration of lead vacancies V_{Pb}'' (see Sect. 1.6). A fairly well written overview about the effects of dopants in the PZT system is given by Xu (1991).

The ionic radii of the relevant dopants in this study are listed in Table 1.1. The oxygen ion O^{2-} has radius 126 pm. Lead has so far only been observed in charge state Pb^{2+} or as a shallow hole trap Pb^{3+}. Warren et al. (1996a) found by ESR that Ti^{3+} will also exist in PZT under certain circumstances (see Sect. 1.6.3).

The dopants lanthanum, nickel and antimony were used in this investigation. The defect reaction

$$La_2O_3 + 3PbTiO_3 \rightleftharpoons 2\left(La_{Pb}^{\bullet}\right)TiO_3 + (V_{Pb}'')\,TiO_3 + 3PbO \quad (1.12)$$

determines the incorporation of lanthanum, and

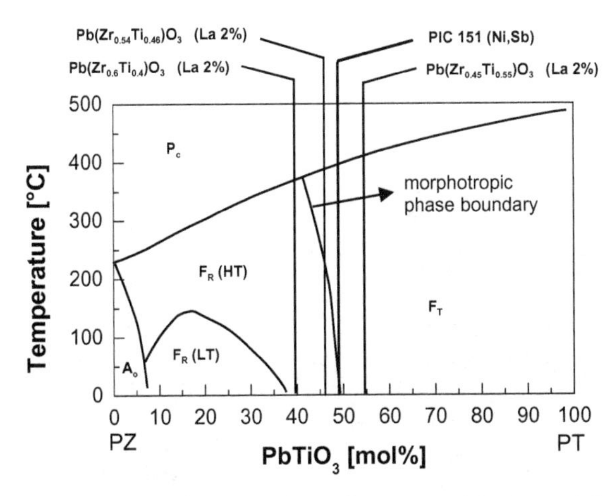

Fig. 1.2. PZT phase diagram including the compositions used for the La-doped PZT and PIC 151. P_c = cubic paraelectric; F_R (HT) = high temperature rhombohedral ferroelectric; F_R (LT) = low temperature rhombohedral ferroelectric; F_T = tetragonal ferroelectric; A = orthorhombic antiferroelectric

$$Sb_2O_5 + PbTiO_3 \rightleftharpoons (V''_{Pb})\left(Sb^{\bullet}_{Ti}\right)O_3 + Pb\left(Sb^{\bullet}_{Ti}\right)O_3 + TiO_2 \quad (1.13)$$

the incorporation of antimony. In the first case lead oxide is liberated during sintering, while in the second case excess lead oxide has to be present to avoid the formation of TiO_2. Titanium was used symbolically for either titanium or zirconium.

Similarly for the acceptor doping

$$NiO + PbTiO_3 \rightleftharpoons Pb\left(Ni''_{Ti}\right)\left(V^{\bullet\bullet}_{O}\right)O_2 + TiO_2 \quad (1.14)$$

excess lead oxide has to be present during processing to avoid the formation of TiO_2.

The above defect reactions assume that electronic defects do not develop directly. This is not necessarily the case as will be discussed in more detail in Sect. 1.6.3 and Chap. 5.

Table 1.1. Ionic radii of PZT cations and relevant dopants

Ion	Pb^{2+}	Pb^{4+}	Ti^{4+}	Ti^{3+}	Zr^{4+}	Sb^{5+}	Sb^{3+}	Ni^{2+}	La^{3+}	Y^{3+}
Radius [pm]	124	84	65	76	81	62	76	72	114	95

Table 1.2. Properties of poled PIC 151

ϵ_{33}/ϵ_o	$\tan\delta$ (10^{-3})	k_p	Q_M	d_{33} $(10^{-12}\,m/V)$	T_c $(°C)$
1840	5	0.60	350	300	225

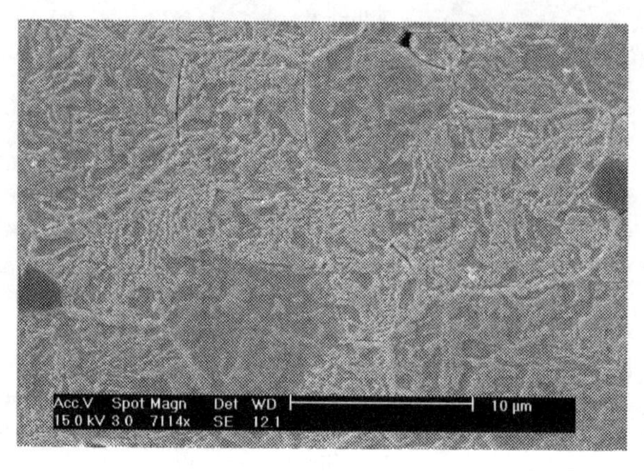

Fig. 1.3. Microstructure of PIC 151 after sintering, etched using HF/HCl

1.5.2 PIC 151

The samples investigated in this study were prepared by the standard mixed oxide route of final composition: $Pb_{0.99}[Zr_{0.45}Ti_{0.47}(Ni_{0.33}Sb_{0.67})_{0.08}]O_3$ = PIC 151 (Helke and Kirsch, 1971). The samples were sintered at 1300°C for 120 min in ambient atmosphere, cut and ground to discs of 10 mm diameter and 0.5 or 1 mm thickness. No fine grained polishing was done on the surfaces.

Silver electrodes were subsequently burnt on for 15 min at 720°C. No further glass additives were contained in the electrode material. Microscopic surface damages in the PZT due to the grinding were healed by the subsequent firing of the electrodes.

Figure 1.3 shows an SEM image of PIC 151 before any cycling was started. Prepoled samples were poled by the manufacturer at 2.5 kV/mm for a few minutes at room temperature. The material properties given by Helke and Kirsch (1971) for $Pb[Zr_{0.44}Ti_{0.48}(Ni_{0.33}Sb_{0.67})_{0.08}]O_3$ are reproduced in Table 1.2. The hysteresis loops including an approximate stress strain loop are shown for reference in Fig. 1.4.

According to the defect reactions (1.13) and (1.14) the sintering of PIC 151 will yield excess TiO_2. This is actually observed in the microstructure, where entire grains of TiO_2 can be found throughout the sample despite sintering in a lead oxide excess atmosphere.

For all samples investigated the geometry of 10 mm in diameter and 1 mm thickness was chosen (unless otherwise specified). To be able to later assign

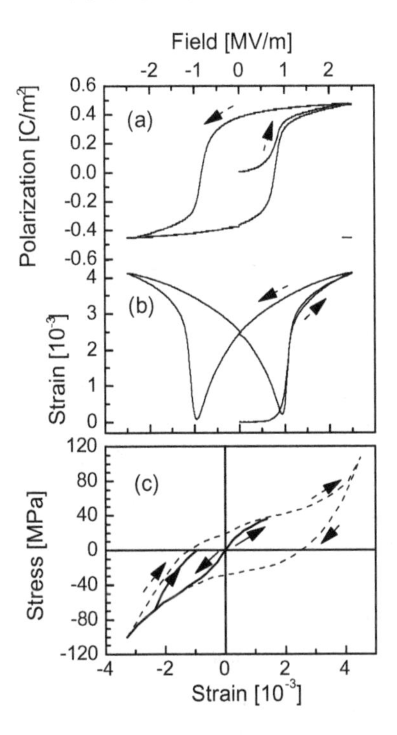

Fig. 1.4. Hysteresis loops of PIC 151. The stress strain data are approximate for the tensile part (dashed lines), stress strain data by Fett et al. (1998)

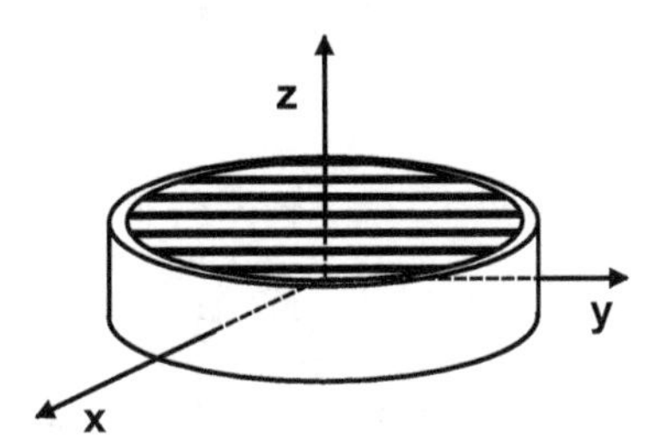

Fig. 1.5. Coordinate system to assign orientations and axes of anisotropy in the samples. The silver electrode is shown along with the 250 μm unelectroded rim

projections and the orientations of anisotropies the following coordinate system is chosen (Fig. 1.5).

1.5.3 Lanthanum Doped PZT

Stoichiometric soft ($Pb_{1-3x/2}La_x(Zr_{1-y}Ti_y)O_3$) PZT ceramics were prepared by a conventional mixed oxide process. The Zr/Ti-ratio was set to 45/55, 54/46 and 60/40 to obtain tetragonal (t), morphotropic (m), and rhombo-

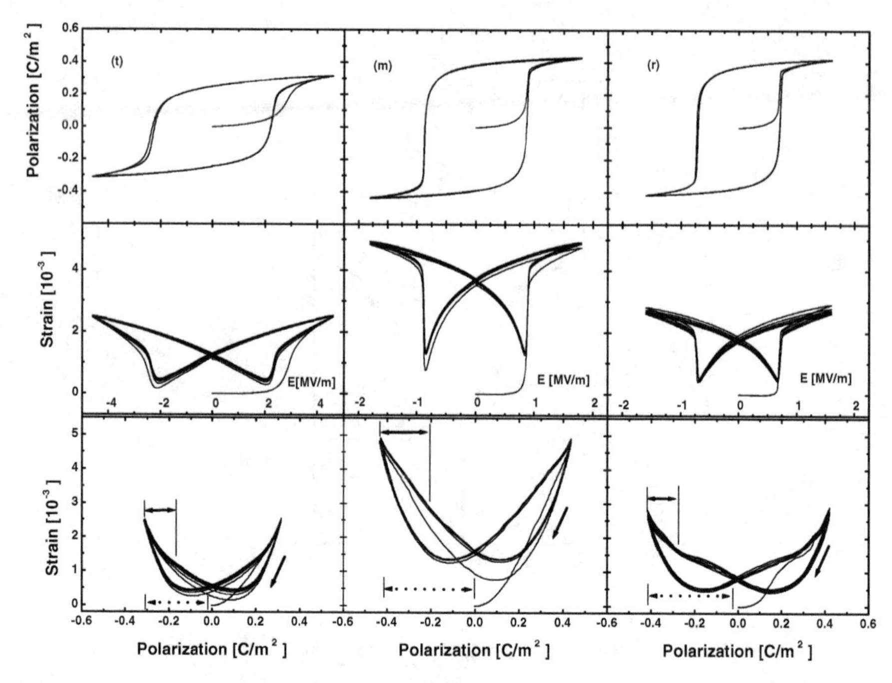

Fig. 1.6. Polarization, strain, and electrostrictive hysteresis for (t) tetragonal (m) morphotropic and (r) rhombohedral coarse grained PLZT (note the different scales for electric field). The double arrows indicate the range of AE-occurrence (see Sect. 4.2.2), the simple arrows the sense in which the electrostrictive hysteresis is traversed

hedral (r) compositions, respectively. The lanthanum content was 2 mole%. Details of powder preparation and the grain size distribution can be found in Hammer (1996); Hammer et al. (1999). For the fine grained (f.g.) and coarse grained (c.g.) compositions the sintering temperatures were 1225°C (c.g.: 1300°C) for 2h (c.g.: 10h). The average grain sizes were: (t, f.g.) 1.6 µm, (t, c.g.): 3.0 µm, (m, f.g.): 1.9 µm, (m, c.g.): 2.9 µm and (r, f.g.): 1.8 µm, (r, c.g.): 3.3 µm. Discs of diameter 10 mm and thickness 0.94 mm were sliced using a diamond saw, polished down to a 3 µm finish and fully electroded with sputtered Au/Pd-electrodes.

Hysteresis loops are shown in Fig. 1.6 for the coarse grained PZT compositions including the virgin curves also termed poling. For tetragonal PZT the inflexion point during poling is distinctly higher than the coercive field E_c in the subsequent cycles, in morphotropic PZT the two are essentially equal, and in the rhombohedral PZT the inflexion field for poling lies beneath the coercive field. The strain data show a similar behavior. The electrostrictive hysteresis (row 3 in Fig. 1.6) shows the known double parabolic shape (Jaffe

Table 1.3. Strain and polarization parameters for coarse and fine grained PLZT (determined at 20mHz). (Δs = strain, p-p means between minimum value at E_c and at maximum field, fg = fine grain, cg = coarse grain)

Quantity		tetra-gonal		morpho-tropic		rhombo-hedral	
		fg	cg	fg	cg	fg	cg
from electric polarization P(E):							
- inflexion point for poling	[V/mm]	2730	2730	940	880	790	690
- coercive field, first inversion cycle	[V/mm]	2430	2300	960	880	880	720
- coercive field, subseq. cycles = E_c	[V/mm]	2330	2200	960	860	860	700
- remnant polarization P_r	[C/m^2]	0.30	0.25	0.39	0.39	0.39	0.37
from strain hysteresis s (E):							
- inflexion point for poling	[V/mm]	2780	2890	950	880	790	690
- minimum, first field inversion	[V/mm]	2280	2130	940	850	860	700
- inflexion, point first field inversion	[V/mm]	2480	2500	1000	900	890	730
- minimum after several cycles	[V/mm]	2200	2030	930	850	840	670
- inflexion point after several cycles	[V/mm]	2410	2330	980	890	870	720
- absolute remnant strain	[10^{-3}]	1.4	1.2	3.5	3.7	(2.5)	1.9
- strain p-p (E_{max}=1.7 E_c)	[10^{-3}]	2.4	2.1	4.2	4.1	2.2	2.2
- strain p-p (E_{max}=4.6 kV/mm)	[10^{-3}]	2.7	2.5	4.8	4.8	(3.7)	2.7
- d_{33} ($E = 0$, ν=20mHz)	[10^{-12} m/V]	390	350	1240	1130	920	1090
- Δs between P_r and P_{max} (E_{max}=4.6 kV/mm)	[10^{-3}]	1.31	1.19	2.33	2.09	1.84	2.00
- Δs between $P(E_c)$ and P_{max} (E=4.6 kV/mm)	[10^{-3}]	2.36	2.0	4.9	5.3	3.5	3.6

et al., 1971; Zorn et al., 1985; Wersing, 1974), again with the first few cycles different. The material thus relaxes into a state of symmetric hysteresis during the first few cycles. The electrostrictive hysteresis is undulated for increasing fields particularly in the rhombohedral composition indicating a range of predominantly 180° switching accompanied by less strain changes than the 70°/110° switching and marking the distinct deviation from purely quadratic electrostrictive behavior. The fine grained materials behave similarly (not shown), but with a less broad electrostrictive hysteresis. It is known for $BaTiO_3$ ceramics that larger grain sizes will reach higher polarization values before strain changes set in, thus the parabola becomes wider at the bottom (Arlt, 1990a). The values for the different hysteresis parameters are given in Table 1.3. The coercive fields depend on grain size in the rhombohedral composition and somewhat in the tetragonal composition. The coercive field of the tetragonal composition is strongly frequency dependent (Hoffmann et al. (2001), see Fig. 2.13). Thus, values from Table 1.3 have to be compared to strain data taken at the identical rate.

1.6 Defects in PZT

The fatigue mechanism in PZT is intimately related to the point defects in the material. This fact is generally accepted now and will be further supported by the work presented here. In order to provide a reference for the later chapters, the defect balances are reviewed in this section.

1.6.1 Vacancies and Interstitials

The most commonly encountered Schottky defect in PZT stems from the liberation of PbO from the matrix material due to its high volatility,

$$\mathrm{O_O} + \mathrm{Pb_{Pb}} \rightleftharpoons (\mathrm{PbO})_\mathrm{g} + \mathrm{V_O^{\bullet\bullet}} + \mathrm{V_{Pb}''} \tag{1.15}$$

yielding PbO vapor in the atmosphere and TiO_2 in the ceramic (case of $PbTiO_3$). Depending on the cross section for recombination of the electronic charge carriers, these will annihilate

$$\mathrm{e}' + \mathrm{h}^\bullet \rightleftharpoons \mathrm{nil} \ , \tag{1.16}$$

or be trapped as localized electron or polaron states (see Sects. 1.6.3 and 5.3.3). If any of the two vapor partial pressures $p_{\mathrm{O_2}}$ or p_{Pb} is enhanced with respect to the other, the reactions (1.17) and (1.18) furthermore enter the defect balance.

$$\mathrm{O_O} \rightleftharpoons \frac{1}{2}(\mathrm{O_2})_\mathrm{g} + \mathrm{V_O^{\bullet\bullet}} + 2\mathrm{e}' \tag{1.17}$$

$$\mathrm{Pb_{Pb}} \rightleftharpoons (\mathrm{Pb})_\mathrm{g} + \mathrm{V_{Pb}''} + 2\mathrm{h}^\bullet \ . \tag{1.18}$$

Neither significant numbers of Ti-vacancies nor interstitials have so far been reported in literature (Smyth, 1985).

1.6.2 Thermal Equilibrium

As all mechanisms involving point defects are thermally excited, crucial differences result at different temperatures. At temperatures well above 700 K the point defects are considered in thermal equilibrium. At lower temperatures some or all defects may be quenched, i.e. will become located to a variable degree. Particularly beneath the Curie points of most common perovskite type ferroelectrics the high temperature equilibria of defects are quenched and inscribed into the material. They will only change towards the low temperature equilibrium during extended time intervals.

$BaTiO_3$ and $SrTiO_3$

The thermal equilibrium conductivity of perovskite ferroelectrics has been studied in a series of articles from the 70's to the 90's. $SrTiO_3$, $BaTiO_3$, as well as all the lead containing compositions like PZT exhibit p-type conductivity at ambient oxygen partial pressure ($2 \cdot 10^4$ to 10^5 Pa) and become n-type conductors at lower oxygen partial pressures.[1] Under normal conditions an ionic crystal should show the lowest conductivity, when it is completely stoichiometric. In the perovskites this is expected for processing under pure oxygen. The behavior exhibited by the commonly manufactured ferroelectric materials is that of acceptor doping. As these materials are mostly produced from reaction grade raw materials, this effect has been ascribed to the natural abundance of lower valent ions brought into the different perovskite type crystals through the minerals or during processing.

For more acceptor doped materials the conductivity minimum at high temperatures shifts further to lower oxygen partial pressures (Kingery et al., 1976). In donor doped ferroelectric perovskites, n-type conductivity is observed for materials processed at an oxygen partial pressure $p_{O_2} = 10^5$ Pa, which is the expected behavior for donor doping (Daniels and Wernicke, 1976).

The predominant defect equilibrium responsible for these effects is the well known incorporation or removal of oxygen from the lattice according to equation (1.17). The corresponding law of mass action is

$$p_{O_2}^{1/2} \cdot [V_O^{\bullet\bullet}] \cdot [e']^2 = K_{O_2} = K_{ox,o} \exp\left[-\frac{\Delta G}{kT}\right] . \qquad (1.19)$$

For intermediate oxygen partial pressures, which is the case for all technologically relevant processing routes, the oxygen vacancy concentration is essentially constant. At identical temperatures a reduction in p_{O_2} induces an increase in electron concentration and vice versa according to the equilibrium (1.19). The equilibrium reaction (1.17) during sintering is crucial for determining the starting defect concentration in a ceramic at room temperature, because the defects are frozen in at about 700 K (Waser et al., 1990a), as already mentioned. Further changes will then take place at a much slower rate and the material stays essentially out of equilibrium (see Sect. 1.6.3).

$PbZr_{1/2}Ti_{1/2}O_3$

Detailed measurements of the defect equilibria in PZT itself are scarce. One of the few analyses was done by Raymond and Smyth (1993, 1994), part of which is also summarized in Waser and Smyth (1996).

[1] (Daniels and Härdtl, 1976; Daniels, 1976; Daniels and Wernicke, 1976; Hennings, 1976; Wernicke, 1976; Eror and Smyth, 1978; Chan et al., 1982; Smyth, 1985; Raymond and Smyth, 1993, 1994; Smyth, 1994; Denk et al., 1995; Maier, 1996)

The crucial difference in defect equilibrium in the lead containing perovskite type ferroelectrics with respect to e.g. $BaTiO_3$ or $SrTiO_3$ is the high volatility of PbO of up to 1.6 to 2%. When thc samples or devices are not sintered in an excess atmosphere of PbO the Schottky defect reaction (1.15) permits a simultaneous occurrence of V''_{Pb} as well as $V_O^{\bullet\bullet}$ at high vacancy concentrations.

Raymond and Smyth (1994) also deduced from their measurements of the Seebeck-coefficient and the constant composition conductivity and constant composition oxygen activity measurements that the conduction in PZT sintered in excess p_{O_2} is that of hopping conduction of holes, eVen at fairly high temperatures. If hopping is already dominant at high temperatures, it eVidently determines the low temperature conductivity contribution of holes to be of the same type (unless variable range hopping becomes dominant). This is true irrespective of the dominant charge carrier, which is the oxygen vacancy at room temperature.

Donors yield excess electrons in the ceramic and will usually shift reactions of type (1.17) towards the left, because the concentration of oxygen vacancies is inversely proportional to the square of the electron concentration. But as long as the number of $V_O^{\bullet\bullet}$ exceeds twice the donor concentration, a significant number of oxygen vacancies survives.

1.6.3 The Quenched State

At temperatures around room temperature the electronic charge carriers are no longer the most mobile defects.

For *acceptor* doped materials a predominant ionic conductivity via the mobile $V_O^{\bullet\bullet}$ and some contribution of p-type electronic conductivity were observed for $SrTiO_3$ (Waser, 1994). The activation energy for the $V_O^{\bullet\bullet}$ motion is 1.0 to 1.1 eV in $BaTiO_3$, which agrees well with high temperature data (Chan et al., 1982). Waser (1994) assigns the mobile electrons or holes to polaron states. They are mobile by a hopping mechanism.

Donor doping yields different conduction mechanisms depending on the microstructure. In single crystals and coarse grained ceramics the conduction is constituted by electrons liberated from the donor centers. In fine grained ceramics, which were sintered in high oxygen partial pressure the cations are completely compensated and the material becomes highly insulating. In both cases there is no contribution from ionic conductivity in the bulk, a limited diffusion of oxygen ions is observed along grain boundaries under gradients of oxygen or electric potential.

The fact that the contribution to conductivity of the vacancies becomes dominant around room temperature can be understood according to a higher energy needed to liberate the holes from their trapping centers. The acceptor ionization energy was determined to be 0.9 eV and the mobility barrier height 0.1 eV in PZT (Raymond and Smyth, 1993), both of which have to be provided, to actually obtain hole contribution to conductivity. The oxygen vacancies need to surmount their mobility barrier.

As the exact nature of the different trapping centers in PZT is not yet known, all the following defect equilibria have to be considered for acceptors A and donors D:

$$A'' \rightleftharpoons A' + e' \tag{1.20}$$

$$A' \rightleftharpoons A^x + e' \tag{1.21}$$

$$D^{\bullet\bullet} \rightleftharpoons D^{\bullet} + h^{\bullet} \tag{1.22}$$

$$D^{\bullet} \rightleftharpoons D^x + h^{\bullet} \;, \tag{1.23}$$

or equivalently:

$$A'' + h^{\bullet} \rightleftharpoons A' \tag{1.24}$$

$$A' + h^{\bullet} \rightleftharpoons A^x \tag{1.25}$$

$$D^{\bullet\bullet} + e' \rightleftharpoons D^{\bullet} \tag{1.26}$$

$$D^{\bullet} + e' \rightleftharpoons D^x \;. \tag{1.27}$$

The obvious acceptors in PIC 151 are nickel and the lead vacancy. The oxygen vacancy and antimony constitute the donors.

Furthermore, Warren et al. (1994b, 1997a) by EPR determined Pb^{3+} and Ti^{3+} to exist in PZT. Thus we have the additional defect reactions

$$Pb_{Pb} + h^{\bullet} \rightleftharpoons Pb^{\bullet}_{Pb} \tag{1.28}$$

$$Ti_{Ti} + e' \rightleftharpoons Ti'_{Ti} \tag{1.29}$$

according to (1.23) and (1.21). Particularly, deep trap states for holes have been found for PZT at about 0.76 and 0.8 eV by Wu and Sayer (1992), and separately at 0.87 eV by Raymond and Smyth (1994) above the valence band. Possible candidates are the defect reaction (1.20) for the lead vacancy or (1.28) for the lead ions, but no unique assignment has been given so far.

Grain Boundaries

Depending on the type of doping grain boundaries may become highly insulating layers or may serve as ionic conductivity paths in other cases. For high acceptor doping in e.g. $SrTiO_3$ the grain boundary itself becomes a highly charged interface layer. The charge at the grain boundary itself is positive. Around the highly positively charged interface a negative space charge layer develops. This layer is highly depleted of oxygen vacancies and holes, while the number of doubly charged Nickel Ni''_{Ti} increases significantly. The layer is around 100 nm thick, smaller than most average grain sizes (Vollmann and Waser, 1994). The grain boundary electronic conductivity becomes four orders of magnitude smaller than the bulk conductivity. Similarly, the grain boundary becomes an almost invincible barrier to the diffusion of oxygen vacancies, which are simply repelled by the positive interface charge. For longer times the conductivity of $SrTiO_3$ becomes strongly field dependent showing a varistor effect. This is due to the high accumulation of charge carriers in the space charge zone around the interface, which in turn induces extremely high potential gradients across the interface. Finally tunneling or electron emission sets in and the current across the boundary rises significantly.

In the donor doped case the sign of all defects is inverse yielding a highly negatively charged interface layer and a positive space charge layer around the interface. The respectively charged ions then accumulate or deplete in the space charge layer (Daniels and Wernicke, 1976).

It is not clear, how far this type of barrier also develops in PZT. As the fatigue behavior in monograin thin films is significantly different from multigrain thin films of similar thickness, such grain boundary layer effects may indeed be the origin of such differences.

1.7 Macroscopic Fatigue

A vast body of literature exists on the fatigue phenomenon.[2] Most of it concerns fatigue in thin films. Different aspects of all these works relevant to our data will be considered along with their presentation in Chap. 2 and discussed in detail in Chap. 5. In the remainder of the introduction general trends will be outlined.

The primary goal for ferroelectric film devices is to maintain a high switchable polarization during many polarization reversals, while in actuator applications the piezoelectric coefficient is intended to stay as high as possible. Even though the polarization in actuators is usually not reversed, same volume parts in a multilayer actuator will experience bipolar or unipolar electrical and compressive mechanical loading at the electrode tips. As most fatal failures of multilayer actuator devices are initiated at these electrode tips, the modifications of the material properties during bipolar cycling have to be thoroughly understood in electrical as well as in electromechanical mixed loading. At the end of this work it will become clear, that despite the fact that only bulk material was investigated, the results are also relevant for thin films.

[2] Scott and Araujo (1989); Scott et al. (1989); Duiker et al. (1990); Pan et al. (1992b); Yoo and Desu (1992c); Desu and Yoo (1993a); Brennan (1993); Furuta and Uchino (1993); Desu and Yoo (1993b); Scott et al. (1993); Mihara et al. (1994b,a); Chen et al. (1994b); Warren et al. (1995b,d); Pawlaczyk et al. (1995); Pan et al. (1996); Gruverman et al. (1996a); Hill et al. (1996); Colla et al. (1997); Warren et al. (1997c); Colla et al. (1998b,c,a); Du and Chen (1998a); Brazier et al. (1999); Weitzing et al. (1999); Kim et al. (1999b); Bobnar et al. (1999); Tagantsev and Stolichnov (1999); Nuffer et al. (2000) and many others.

The primary indication of fatigue is a reduction of the switchable polarization affecting the performance of actuators as well as memories to the same degree (see Fig. 2.2). All referenced articles in footnote 2 start their considerations from this indication of fatigue and correlate it to other modifications of their samples or the external boundary conditions during experiment.

The literature on bipolar fatigue in bulk ceramic lead zirconate titanate (PZT) is by far not as vast as that on thin films. In the detailed discussion in Chap. 5 it will be shown that some aspects are found in both geometries, while a few are only found in either thin films or bulk material.

Fatigue of ceramic PLZT (lanthanum doped PZT) was intensively investigated by Pan et al. (1992b). Their data from PLZT 7/56/44, a composition which is exactly at the morphotropic phase boundary, show the now generally accepted trend, that unipolar cycling yields only a fraction of the fatigue effect of bipolar cycling. Around 15% loss in switchable polarization ΔP were observed after 10^6 cycles unipolar at 10 Hz, while an immediate deterioration down to 30% at $2 \cdot 10^5$ was observed for bipolar cycling (well polished electrode surfaces). The same authors showed that fatigue occurs predominantly in materials exhibiting ferroelectricity and ferroelastic effects. Purely electrostrictive rhombohedral 9/65/35 neither fatigues under bipolar nor unipolar cycling up to 10^6 cycles.

A particular finding by Pan et al. was that fatigue is anisotropic. While polarization and strain degrade in the direction of the applied electric field, the application of electrodes perpendicular to this direction yields an almost unfatigued material behavior (see also Sect. 2.4). Another finding by Pan et al. (1989) was the fact that antiferroelectric compositions show much less degradation than similar ferroelectric compositions in the system $(Pb_{0.97}La_{0.02})(Zr_xTi_ySn_z)$. This fact will also be of relevance later (Sect. 5.8.5).

Along with the polarization, the piezoelectric coefficient drops according to equation (1.6). This has been observed for bulk materials (Pan et al., 1992b) as well as thin films (Colla et al., 1995). In the case of bipolar electric fields well beneath the coercive field, the inverse effect can sometimes be observed, namely increased small signal parameters like d_{33}, k_p, or capacitance Cain and Stewart (1999).

Another effect in ceramic PZT, which is hard to observe in thin films (Bobnar et al., 1999), is the reduction of the strain hysteresis along with the loss in polarization. Weitzing et al. (1999) found simultaneously with us, that besides the reduction of the amplitude of the strain hysteresis loop ("butterfly"), an asymmetry of the butterfly is induced (see Fig. 2.4). The reason for this effect will be given in Sect. 2.2.

In the following sections the different approaches in literature to understand this general material behavior based on some microscopic or mesoscopic mechanisms will be outlined.

1.8 Role of Point Defects in Fatigue

The role of point defects in the fatigue mechanism of ferroelectrics is by now generally accepted. The crucial question of how the fatigue itself arises is still open. Particularly in polycrystalline bulk ceramics the underlying mechanisms seem to be dominated by anything from electronic, ionic, or grain boundary effects to microcracking.

The most localized mechanism for the fatigue phenomenon was discussed by Miura and Tanaka (1996b). The basic idea behind this approach is the change of the binding between different ions of the perovskite unit cell during fatigue. In order to establish the ferroelectric ordering itself, the bond type changes slightly beneath E_c. The crucial difference arises in the titanium-oxygen bonds. Miura and Tanaka (1996b) mainly discuss the influence of the π-bond between the Ti^{4+} and the O^{2-} as directly related to the ferroelectric effect. Molecular orbital calculations of the bond order of this bond show a maximum for a particular displacement of the Ti^{4+}-ion thus inducing the ferroelectric effect. For not fully ionized $V_O^{\bullet}$ in the vicinity of the Ti^{4+}-ion extra electrons are present perturbing the π-bond and thus the source of ferroelectricity on the unit cell level. In a subsequent work the same authors calculated the influence of La^{3+} donor doping on the electronic states of the unit cell. As the electronic La $4f$-states lie energetically between the oxygen $2p$ and titanium $3d$ levels, remaining electrons of a not fully ionized $V_O^{\bullet}$ preferably occupy the La $4f$-states. This permits the above mentioned π-bonds to fully develop and fatigue becomes less probable from a single unit cell perspective. The overall defect chemical charge balance of the crystal is not considered in this work, thus merely the number of oxygen vacancies will determine a fatigue mechanism based on this model.

Park and Chadi (1998) also considered the very local structure around a charged defect. They showed that tail-to-tail domain structures are already induced by single defects in their immediate vicinity. This effect will be discussed in detail along with the fatigue model in Sect. 5.8.5.

General agreement exists about the fact that the number of vacancies or at least their charge state determines the rate of fatigue. The most eVident indication of this phenomenon was given by fairly recent measurements of the fatigue rates in thin films by Brazier et al. (1999) under varying oxygen partial pressure ((111)-oriented film, 300 nm thick, Pt-electrodes). While we ourselves were able to prove, that the exchange of oxygen with external atmospheres was negligible for bulk materials (see Sect. 3.6), thin films have a sufficiently large surface and small thickness to permit an equilibration with the environment. For oxygen partial pressures of around $2 \cdot 10^4$ Pa, which is just the ambient air concentration, the fatigue rate at room temperature was lowest, increasing for higher or lower partial pressures. It is not exactly clear, whether the fatigue is related to the lower amount of electronic charge carriers, or actually the reduced number of vacancies. As the range of oxygen partial pressures investigated was not large (100 Pa to 10^5 Pa), the general defect chemical analysis assumes the number of vacancies to be constant and only their charge state to be changing in equilibrium (Smyth, 1985). As fatigue to a fairly high degree is a non-equilibrium process, the differentiation between the ranges of only charge state changes or increased number of ionic defects is not clear yet.

Similar to an increased p_{O_2}, donor doping improves the fatigue resistance of $BaTiO_3$ to some extent. Nb-doping was shown to shift the logarithmic regime to about one order of magnitude higher cycle numbers for 3% Nb-doping (Chen et al., 1994a). According to the high temperature conductivity data by Daniels and Härdtl (1976) $BaTiO_3$ changes from p-type to n-type conductivity at much lower donor concentrations (already 0.1% La) are sufficient. Again it is not completely clear whether the number of oxygen vacancies is reduced according to equation (1.19), or only the charge state of the vacancies is changed according to (1.20) through (1.27) due to fatigue.

Another study that recognized the relevance of the oxygen partial pressure during thin film processing of PZT was performed by Bernstein et al. (1992). While the fatigue in films sintered in 1 bar of oxygen leveled off, the fatigue in films sintered in 50% oxygen continued logarithmically after 10^8 fatigue cycles. They did not observe a significant influence of lead excess on the fatigue properties.

Extensive studies on the fatigue of ferroelectric materials and the influence of point defects were performed by a group from the Sandia National Laboratories in Albuquerque discussed in the next paragraphs.

A fact that will be picked up in later chapters is their differentiation between strongly clamping and weakly clamping defects, which both reduce the amplitude of the polarization hysteresis. Warren et al. (1994a) investigated Fe-acceptor doped $BaTiO_3$ (single crystal 200 µm thick, Pt and tin-doped In_2O_3 electrodes, ITO) under reducing atmosphere. The domains were pinned by reducing treatment in vacuum (600°C, $p_{O_2}= 10^{-6}$), but the hysteresis could be restored by a bias field and UV light illumination. The number of isolated Fe^{3+}-ions changed considerably under this treatment, but the number of Fe^{3+}-$V_O^{\bullet\bullet}$ pairs was not altered (EPR measurements). They furthermore showed that hole injection at room temperature (UV illumination of the positive electrode, +7 V) reduced the mobility of the domains and reduced the polarization hysteresis, while a subsequent treatment by -50 V re-established the full hysteresis. The number of isolated Fe^{3+}-ions dropped accordingly indicating the formation of Fe^{4+}. In a similar sense the electron injection restored the polarization hysteresis after the reducing atmosphere treatment, while hole injection did not. Thus the reducing condition also recharges the Fe^{3+} to Fe^{4+}, which is contradictory to the anticipated higher electron availability duc to the $V_O^{\bullet\bullet}$ formation. The domain pinning was ascribed to localized electronic charge carriers.

A harsh reduction of the $BaTiO_3$ in 5% H_2 (95% N_2) resulted in a strongly reduced number of Fe^{3+}-$V_O^{\bullet\bullet}$ complexes and P_r was reduced 3 times stronger than in the above case. None of the above treatments resulted in a restoring of the polarization hysteresis nor in an increased number of Fe^{3+}-$V_O^{\bullet\bullet}$ complexes. The Fe^{3+} is considered to be transformed to Fe^{2+} under these strongly reducing conditions, a fact also pointed out by Baiatu et al. (1990) and differently treated in most other defect equilibrium studies. They conclude that domain pinning under highly reducing conditions is constituted by hard defects which are ions or vacancies (see the discussion on hard agglomerates in Sect. 3.1).

In a further study the same group could show that UV light illumination can restore a full hysteresis in PLZT, which had been pinned by defects introduced during processing (Dimos et al., 1993). The stretched exponential time dependence of restoration of the full hysteresis indicated several trapping states for electronic charge carriers. Furthermore, the existence of Pb^{3+} (and Ti^{3+}, but this signal was very weak) was shown by EPR measurements. UV illumination (2.3 eV) allowed to empty these defect states, but the hysteresis did not improve in amplitude. Thus Pb^{3+} is not a pinning center for domains in PLZT. The formation of paired electronic defects in PLZT is suggested, because no paramagnetic centers obviously related to the reduced hysteresis amplitude could be assigned by EPR measurements.

The data on the domain trapping by electronic or ionic charge carriers was correlated with fatigue data on PZT thin films (Warren et al., 1995c, 1997c). They observed that the Fe^{3+}-$V_O^{\bullet\bullet}$ complex changes its character during fatigue and the coupling between the magnetically sensitive Fe^{3+}-ion and the $V_O^{\bullet\bullet}$ is changed. As 70% of the g-factor is determined by the next nearest neighbor the changes in g-factor were ascribed to changes of the octahedron about the Fe^{3+}-ion (see discussion of this point in Sect. 5.8.2).

In thin PZT films UV light and bias fields close to the coercive field in the direction opposite to the poling field yield the highest imprint and strongest loss in polarization (Warren et al., 1995d,c). The imprint due to fatigue is significantly different. Their major result concerning fatigue is that the fatigue induced polarization suppression can neither be removed solely by reverse saturating fields nor by UV irradiation alone, which is different from the cases, when the imprint is induced due to the field/UV treatment. Only their combination allows recovery of the polarization lost during fatigue. Similarly a combination of heat treatment (100°C) and a bias field yield some restoration of the polarization.

Another result mentioned in the same publication is the fact that a small part of the total polarization (10%) could not be restored by any treatment (UV + electric field, UV + temperature). This points to a mechanism establishing very sTable defect agglomerates, because microcracking is unlikely to be a cause in thin films and has not been reported so far.

Lee et al. (1995a) showed in a similar study the direct correlation between UV light illumination (He-Cd, 325 nm $\hat{=}$ 3.8 eV) and the fatigue of PZT. Under UV light illumination PZT films of 500 nm thickness with 30 nm

- 40 nm thick Pt-electrodes on the light incident side and YBCO-electrodes as backing exhibited retarded fatigue. The unfatigued hysteresis cycle displays slightly larger remnant polarization, but the effect immediately ceases when illumination stops. The higher the illumination during fatigue the less the polarization drops with cycle number. Still, the onset of fatigue occurs at about the same cycle number of $5 \cdot 10^6$ cycles. The difference between the illuminated and non-illuminated remnant polarization rises exponentially to 15% at 10^9 cycles, when the non-illuminated remnant polarization has dropped to about 60% of its initial value. The photocurrents were very small in unfatigued samples and only displayed the edge current peaks due to pyroelectricity, while in the fatigued state the photocurrent was much higher for negative bias (on Pt) and actually of the same sign eVen under positive bias, which reflects the different work functions of the electrodes.

1.9 Domain Pinning, Aging, and Imprint

Aging has to be strictly separated from fatigue. Aging is the change of material properties due to time *without* changes of the external boundary conditions, while fatigue is the degradation of material properties *due to* cyclic changes of the external boundary conditions. Even though some of the microscopic mechanisms may be similar, the notations have to be properly used.

Aging was phenomenologically discovered early in the studies of $BaTiO_3$ by McQuarrie (1953) and Plessner (1956). Soon a list of dopants was available, which induce aging (McQuarrie and Buessem, 1955). Jonker (1972) then found that in the vicinity of phase transitions aging at different temperatures can reversibly induce crystallographic phase changes. Later, aging was associated with a microscopic bulk effect in $BaTiO_3$ with Mn^{3+}-$V_O^{\bullet\bullet}$ complexes being the stabilizing entity (Lambeck and Jonker, 1978, 1986). Depending on the oxygen partial pressure applied at high temperatures, the Mn would take valence Mn^{3+} or Mn^{4+}. Only those crystals with significant amounts of Mn^{3+}-$V_O^{\bullet\bullet}$ complexes showed the aging effect. Depending on whether c- or a-domains were stabilized, the hysteresis showed a bias field for stabilized a-domains or a double hysteresis like in antiferroelectrics for stabilized c-domains. A distribution of relaxation rates

$$E_i(t) = \sum_{\gamma} E_{i0,\gamma} \left(1 - e^{-t/\tau_\gamma}\right) \tag{1.30}$$

was determined to model the relaxation behavior into the aged state. Depending on the aging times, the velocities of the domain walls were determined according to the model for 180° domain wall motion by Miller and Weinreich (1960). These experiments clarified the bulk character of the aging effect.

Simultaneously, the group around Arlt developed a model of aging due to the reorientation of defect-dipoles in hard doped ceramics. The formation of acceptor-$V_O^{\bullet\bullet}$ pairs and the charge states of the acceptor dopants were early determined by ESR measurements, Mößbauer spectroscopy and optical absorption measurements (Hagemann, 1980).

The development of the offset-field was modelled in the framework of the orientation of defect dipoles (Arlt and Neumann, 1988). The underlying mechanism is based on the fact that a high degree of association occurs between acceptor dopants and oxygen vacancies (see also p. 119). These pairs form atomic dipoles in single unit cells. Under the assumption of immobile acceptor dopants and a significantly higher mobility of the oxygen vacancy Arlt and Neumann (1988) differentiated between the different possible positions of the oxygen vacancy, when the defect pair is embedded in a domain of a particular polarization direction. The transition energies for jumps between the three possible orientations of the dipole with respect to the surrounding domain were in the range from 0.03 to 0.7 eV depending on the way they were calculated. A Landau-potential was used for the Gibb's function with an additional linear term in the spontaneous polarization times the internal bias field. The bias field in turn was determined by the jump rates between the six neighboring oxygen sites around the Ni dopant ion. For long times an equilibrium population of the non equivalent neighbor sites is reached, which lowers the total energy of the domain. Upon domain reversal, this energy is added to the total energy and the ferroelectric hysteresis is shifted along the electric field axis by a bias-field. In the same spirit a later work of this group described the reduction of the internal bias field due to the application of bipolar external fields (Lohkämper et al., 1990). The introduction of a small distribution of relaxation times of Gaussian form rendered their model suiTable to describe Ni-doped $BaTiO_3$. Similar to the high field measurements the small signal dielectric constant is altered due to aging. Particularly the extrinsic contributions to ϵ due to the 90° domain walls is reduced, which became eVident in the imaginary part of the dielectric constant (Robels et al., 1995a).

Different from the assumptions by Arlt and Neumann (1988) and Robels et al. (1995a), Chen and Fang (1998) ascribe the offsets in $C-V$-measurements (capacitance-voltage) after d.c. field application to the injection of electrons and their stable localization within PZT thin film capacitors (PZT 53/47, 300 nm, Pt-electrodes). The electrons become localized at charged domain boundaries. They determined an electron trap density of $1.2 \cdot 10^{13}\,\text{cm}^{-2}$ and an effective cross section for electron capture of $1.9 \cdot 10^{-19}\,\text{cm}^2$.

Along the lines of the defect orientation worked out by Neumann and Arlt (1986b), Warren et al. (1995a) used the EPR technique to show the alignment of the Fe^{3+}-$V_O^{\bullet\bullet}$ defect dipoles under a d.c. bias at elevated temperatures in polycrystalline $BaTiO_3$ (-450 V, 110°C, 1 mm thick). They proved a very high alignment of the dipoles along the polarization of the ceramic confirming the model assumptions by Neumann and Arlt (1986b). An interesting fact about their study is the sensitivity of the material to UV light. Under UV light illumination and d.c. bias the EPR signal of the Fe^{3+}-$V_O^{\bullet\bullet}$ defect dipoles increased by a factor of 10 during 2000 s, while in the non-illuminated case the signal eVen drops. A second defect dipole, which they could not definitely assign to a microscopic source ($g_\perp^e = 3.955$, $S = 3/2$, also highly aligned and high tetragonal distortion, possibly Fe^{1+}-$V_O^{\bullet\bullet}$ or Fe^{5+}-V_{Ba}'') slightly dropped in signal for the non-illuminated sample and strongly dropped for the UV case opposite to the Fe^{3+}-$V_O^{\bullet\bullet}$-signal in the second case.

The actual orientation of the inherent defect dipoles was determined later (Warren et al., 1997b). Again electron paramagnetic resonance (EPR) measurements were performed on samples aged using an accelerated aging process at higher temperature. The measurements revealed, that Fe-doping will yield Fe-$V_O^{\bullet\bullet}$ complexes in $BaTiO_3$, if a sample is held in an electric field at elevated temperature, which is equivalent to the local fields exerted by the polarization of the domain. An effective bias field is established. In a further EPR study, the alignment of the $V_O^{\bullet\bullet}$-Fe^{3+} -complexes along the local polarization of the domain (Warren et al., 1996a) was explicitly shown confirming the earlier bulk assumptions (Robels and Arlt, 1993). The orientation of the defect dipoles does not occur above the Curie temperature nor in non-polar perovskites like $SrTiO_3$. In this work they also discuss the particular role of Pb in perovskite compounds. They assigned parts of the high polarizability in PZT to the covalent character of the Pb-O bond. half the bonds are slightly shorter indicating a possible contribution of Pb^{2+} to the ferroelectric polarization switching beyond the standard displacement of the Ti^{4+}-ion (see also Miura and Tanaka (1996b), p. 20).

A different perspective on aging data resulted from another data set from the Sandia group (Dimos et al., 1994; Warren et al., 1996b). Under UV illumination at band gap energy PZT as well as $BaTiO_3$ could be strongly aged. They assigned this aging phenomenon to the excitation of electronic charge carriers and their subsequent accumulation at domain walls. A similar effect can be achieved by heating a poled sample near the Curie point, but still sufficiently low to be in the ferroelectric phase. By simple heating it is on the other hand not clear that only electronic charge carriers accumulate and subsequently pin domains. This will be discussed in some more detail in Sects. 5.4.2 and 5.4.4. A major influence with respect to these effects stems from the microstructure. The externally provided charges on the electrodes are sufficient to compensate the effective externally experienced polarization. Locally the polarizations are not fully compensated, because of the arbitrary orientation of the grains with respect to the external field direction, unless mobile charge carriers are present in the microstructure. Exactly these charge carriers are provided by the UV light and accumulate locally to fully compensate all depolarizing fields due to the spontaneous polarization. If a bipolar field is applied after such an aging procedure, an offset field is observed in the polarization hysteresis. This type of aging is termed imprint in thin film technology. Thermally imprinted offsets are usually more stable than the UV imprinted ones (Pike et al., 1995; Warren et al., 1995e). Thus a significant contribution from ionic defects is anticipated for thermal imprint. Dipole alignment is very high under high electric fields as described in the model by Robels et al. (1995a).

On the other hand, and this is a crucial statement with respect to shelf aging as well as fatigue, the alignment of defect dipoles at remnant polarization is low without applied external fields, some increase in temperature or UV illumination as was proven by Warren et al. (1995a), again using EPR-measurements. Thus the lower values of the reorientation energies determined by Arlt and Neumann (1988) are not suitable, because realignment will not happen at room temperature, while the slightly higher values seem very reasonable.

A brief summary of the work of the Sandia group is given by Warren et al. (1996b).

Fernandez et al. (1990) report a study on the effects of aging on conductivity in undoped and Nb-doped $BaTiO_3$. While in as-sintered samples the conductivity of the grain boundaries is lower by about an order of magnitude, this difference is very much reduced after 1 year of shelf aging. The authors attribute this effect to the aging induced segregation of niobium into the grain boundaries, where they enhance the conductivity, while in the bulk the conductivity is reduced. This also demonstrates the close correlation between the mobile species in the bulk with the dopant ions. The influence of grain boundaries will be further discussed in Sect. 5.7.

In most actuator applications the driving voltage will not be bipolar or at least not exceed the negative coercive field. Particularly when trying to obtain as high strains as possible by driving the actuator somewhat into the negative polarity, a fatigue mechanism is observed, which is distinctly different from those that will be the major subject of this study. It actually concerns the partial depoling due to a loss of remnant polarization with cycle number. A recent work by Polcawich and Trolier-McKinstry (2000) investigated the possibility to improve the fatigue resistance in this sense for micromechanical applications. UV illumination or thermal poling were able to generate a sufficiently strong imprint in the material. Bipolar fields of about 0.5 times the coercive field did not depole the sample any more. The imprint by UV illumination was more sTable than the thermal one, which is actually opposite to the results found by Pike et al. (1995) and Warren et al. (1995e) as mentioned above. Thus, it seems to depend on the particular material, whether the UV or thermal treatments generate more sTable offsets. Pure aging (shelf aging) after UV illumination actually increased the net piezoelectric coefficient d_{31} with time in the measurements by Polcawich and Trolier-McKinstry (2000). All imprint treatments on the samples resulted in an effective offset field.

In a study using very high purity $BaTiO_3$ Wu and Schulze (1992a,b) determined the influence of aging on the different parts of the dielectric permittivity and dielectric losses. The dielectric permittivity of a ferroelectric is constituted by a bulk and a domain wall contribution. By modifying the a.c. field level, the authors could show, that the aging hardly influenced the low amplitude dielectric measurements, which are sensitive to the bulk permittivity. At higher measurement a.c. field levels the well known loss of dielectric constant after aging was observed along with a reduction in the dielectric losses. The authors thus ascribe the aging mechanism to some point defect reordering, that reduces the domain wall mobility. Bulk scenarios for the blocking of domains are not as relevant in their opinion.

A two dimensional finite element model for describing the effect of local domain pinning on the macroscopic fatigue phenomenon was developed by Wang et al. (1996). The authors conclude that a simple criterion for domain pinning is sufficient to infer the reduction of the macroscopic polarization. Thus any local mechanism providing some pinning will macroscopically be reflected in the same way as was also discussed by Warren et al. (1996b).

1.10 Role of Electrodes and Thin Film Fatigue

The influence of electrodes on the ferroelectric properties is much more pronounced in thin films than in bulk ferroelectrics. Only those topics directly relevant to bulk fatigue will be presented here. A more detailed summary of the topic can be found in the book by Scott (2000).

Metal Electrodes

Two major microscopic mechanisms occur at metal electrodes. On the one hand, electronic space charges are formed underneath the electrodes forming a Schottky barrier due to the semiconducting character of the ferroelectric. On the other hand, the ion diffusion of certain ionic species, particularly oxygen, is very limited in metals and an increase or decrease in concentration of certain ionic species, particularly oxygen vacancies, occurs near the electrodes. As these defects are charged, they form an ionic space charge layer. The differentiation between both types of space charge layers is difficult, but will nevertheless become important when discussing the different fatigue mechanisms in Chap. 5.

One of the first works to discuss space charge layers at the surface of a ferroelectric crystal was Känzig (1955). The space charge layer is formed by defects introduced by the external atmosphere, later mostly determined as $V_O^{\bullet\bullet}$. This surface layer was observed to generate fields in the crystal high enough to maintain the tetragonal distortion of $BaTiO_3$ well above the Curie temperature, which was verified by X-ray and electron diffraction data. Thicknesses of these layers range from 10 to 1000 nm and induce fields on the order of 1 - 100 kV/mm. Fairly high $V_O^{\bullet\bullet}$-concentrations $> 10^{17}$ cm^{-3} are present within the layers.

Triebwasser (1960) then determined the polarization hysteresis while applying external dc-bias fields. While the polarization hysteresis displays normal behavior, the field dependent dielectric constant values become considerably asymmetric after a long time dc-field application. The effect is explained similarly to Känzig's work by a simple Schottky barrier forming underneath the electrodes.

During the early studies of fatigue in ferroelectrics the electroding metals were varied (Teowee et al., 1995). It was found that indium electrodes showed the least degradation (Fraser and Maldonado, 1970). Mehta et al. (1973) then investigated the effect of electrodes on the retention failure of ferroelectric memories. They discuss a scenario in which the screening length of the polarization differs between metal and semiconducting electrodes. In the case of metal electrodes the depolarizing field originating from the ferroelectric itself is not sufficiently screened by charges in the electrodes. They assume this depolarizing field to persist eVen in a switched memory and consider it the origin of the retention loss in a ferroelectric memory cell. They used a model that actually had to assume very low coercive fields of the thin film ferroelectric to explain the data of depoling with time in the memory cells. The effect was more severe in thin than thick films.

Brennan (1995) used a time independent electrochemical approach to describe the effect of donors and acceptors in the proximity of the electrodes. A space charge layer is formed underneath the electrodes. Under the influence of atmospheric oxygen and ionization energies of around 1 eV for the naturally occurring acceptors in perovskite ferroelectrics like PZT, only partial ionization of the acceptors is achieved without an electric field present. The response of the partially ionized defects to the externally applied electric fields is considered the fatigue mechanism. The contact potentials to the metal electrodes will further modify the defect concentration by lowering the $V_O^{\bullet\bullet}$-concentration in case of a negative contact potential and increasing it for a positive contact potential. A depleted n-type semiconducting layer constitutes the space charge layer underneath the electrodes. The Schottky barrier models are supported by this point defect model approach. A crucial result of the calculations is the fact that the acceptors are not fully ionized at room temperature and atmospheric oxygen partial pressures. The relative difference in concentration of $V_O^{\bullet\bullet}$ with respect to $[A']$ will determine, whether the material becomes p or n-type and this difference may be induced by the external electric field.

Under the application of an electric field, drift currents and electronic currents are induced. Poisson's equation then determines the local fields, while charge neutrality and the electron-hole equilibrium are no longer applicable according to Brennan (1995), the latter point being treated very differently by Waser and Hagenbeck (2000). If the mobility of $V_O^{\bullet\bullet}$ is considered low in the short time limit, the concentration of charged acceptors [A′] near an electrode of positive potential is highly enhanced in a layer about 200 nm thick, and slightly lowered in the inverse case. For longer times, the charged acceptor concentration remains similarly high at the electrode but the $V_O^{\bullet\bullet}$ concentration drops at the electrode at positive potential. The inverse case induces a very high increase of $V_O^{\bullet\bullet}$ at the negative electrode. The accumulated $V_O^{\bullet\bullet}$ correspond to a chemical reduction of the layer beneath the electrode and vice versa. The contact potential can equally well be the source of the local field as the external field. Different from assumptions by Baiatu et al. (1990), Brennan does not assume local equilibrium, because the generation/recombination rate is very much lower than in ordinary semiconductors. Furthermore, the mass action between holes and electrons has to be ignored in such poor conductors under externally applied fields.

Stolichnov et al. (1999) also determined the formation of a space charge layer during fatigue of PZT 45/55 thin films (450 nm, Pt electrodes) permitting the tunneling of electrons by leakage current measurements (E = 300 kV/mm) at low temperatures (100-140 K). The samples were fatigued at 25 kHz at room temperature (square bipolar, cycle number not specified). The temperature independent current-voltage characteristics confirm a tunneling conduction mechanism. The tunneling is considered to occur from the PZT valence band to the platinum Fermi level. The necessary band bending at the interface implies that PZT is a p-type hole conductor. a.c. cycling does not induce a significant change in the current characteristic before the polarization switching is significantly suppressed. Heating up to 490 K, 10 minutes, restored the original conduction, but only partially restored the polarization suppression. Trapped charge at the electrode interface is made responsible for the fatigue. For this to occur, charges of the opposite sign than those which are tunneling have to constitute the space charge, thus in this case trapped electrons (e.g. polarons, see Sect. 5.3.3). A near electrode layer of trapped electrons in the PZT fits a tunneling model, which they used. As the recovery of polarization at 490 K is only partial, the authors allow at least two fatigue mechanisms to exist simultaneously.

The formation of a space charge layer beneath metal electrodes (Au) was directly shown by Kundu and Lee (2000). The authors prepared thin films in the range from 70 nm to 680 nm and measured the dielectric constant, polarization, and leakage current. For a 70 nm thick film the remnant polarization vanishes almost entirely, but already films of thickness 130 nm show approximately the normal remnant saturation and 60% of the remnant polarization, but very high coercive fields (2× and 3× the value of the thicker films). The measurement of leakage current demonstrated, that the conduction is grain boundary limited in films of thickness larger than 300 nm showing varistor type behavior at low voltages (grain size is about 100 nm) and ohmic current beneath. At high fields (> 20 kV/mm) the latter samples showed space charge limited current, while the former showed breakdown. Thus, a thickness of a most likely ionic space charge layer of around 50 nm exists in such films. The conduction in the grains is dominated by deep traps, while in thicker samples the grain boundaries parallel to the external electrodes are current limiting.

Microscopic mechanisms of fatigue correlated to electrodes as starting point of dendritic growth of semiconducting paths into the ferroelectric were discussed by Duiker et al. (1990) for thin films (details see p. 38).

Electrode Quality

A series of studies showed the relevance of a tight connection between the electrode and the ferroelectric material. If this intimate connection is not guaranteed, fatigue will start at some local flaws in the electrodes, which serve as field and stress singularities.

Bipolar cycling with rough electrodes was shown to lead to much earlier degradation in a set of $(Pb_{0.97}La_{0.02})(Zr_xTi_ySn_z)$, $x = 0.60-0.66, y = 0.09-0.12, z = 0.23-0.35$ ferroelectric to antiferroelectric ceramics (Pan et al., 1989). In a work along the same lines Jiang et al. (1994a) investigated the quality of electrodes and found that evaporated gold electrodes lead to much higher degradation than sputtered ones. Cleaned PLZT 7/68/32-samples for which all solvents were burnt out at 500° - 600°C did not show any fatigue despite the fact that the surfaces were not polished (Jiang et al., 1992). Thus, a clean electrode contact is crucial irrespective of further effects of the electrode material on fatigue.

A different aspect of electrode processing was addressed by Thakoor (1994). They discuss the annealing treatment of lift-off patterned Pt-electrodes after deposition onto thin film PZT ferroelectric thin films. An annealing treatment of 560°C after deposition best improves the switching charge stability. The tests were performed as consecutive voltage pulses of identical polarity. The lower the secondary charging current, the better the device performance. Additionally, a photoconductivity study was performed each after 60 s to ensure that only bulk and not space-charge-affected currents occur. Well annealed samples show constant photocurrent irrespective of bipolar cycle number. The improvement is assigned to a reduced formation of an interface layer beneath the electrodes. The photocurrent response is interpreted as the net photovoltaic current of two back to back Schottky diodes (Thakoor and Maserjian, 1994).

Oxide Electrodes

A series of oxide electrodes has been tested for PZT, particularly in thin films. It was found that all electrodes permitting oxygen diffusion improved the fatigue resistance of the ferroelectric by several decades in cycle number.[3]

[3] RuO_x, Bernstein et al. (1993); Vijay and Desu (1993); Bursill et al. (1994); Lee et al. (1995b); Desu (1995); Taylor et al. (1995); Kim et al. (1999a); $La_{0.5}Sr_{0.5}CoO_3$, Ramesh and Keramidas (1995); Al-Shareef et al. (1996b); Cille-

The common understanding of this effect is that due to the very low ionic conductivity across a metal, a layer of ionic defects forms underneath the metal electrodes. This layer displays a reduced dielectric constant. The potential difference will therefore generate a higher electric field in these low-ϵ regions near both electrodes and effectively reduce the potential drop across the ferroelectric. As in thin films the hysteresis loop is very slim and slanted as compared to bulk material, an effectively much lower maximum polarization is reached at the identical external potential difference compared to an unfatigued sample. The blocking layers have finite thickness and thus thicker samples suffer less from the influence of the blocking layer (Lee et al., 1995b). Another effect of the interface layer beneath the electrodes is the bending of the band structure in a similar sense as for Schottky diodes in semiconductors. This will be further discussed in Sect. 5.6.

Different from all other studies mentioned so far Lin et al. (1997) found some fatigue in films with rhombohedral crystal structure on oxide electrodes (PLZT 3/54/46 and 3/34/66). They observed the onset of degradation at about 10^7 cycles.

Semiconductor Electrodes

In an earlier work on the influence of electrodes on the polarizability of TGS (Trigylcine Sulfate) the influence of semiconducting electrodes on the formation of a space charge was modelled by Wurfel and Batra (1974). They showed that for a highly donor doped semiconducting electrode material the TGS sample had to reach 1 mm thickness to be able to develop the spontaneous polarization of the free sample. Beneath this thickness the space charge layer formed at the electrode was strong enough to generate a severe depolarizing field.

A study of the effects of semiconductor electrodes on the behavior of several oxide ferroelectrics was provided by Xu et al. (1990). Their study showed that depending on the particular ferroelectric substance, p-n-junctions are formed at the n- or p-doped silicon electrodes. Strongly rectifying characteristics were obtained for PZT (0.15% Nb) , $BaTiO_3$, SBN and PBN on n-type silicon and $KNbO_3$ on p-type silicon and weakly rectifying on inversely doped silicon. Only SBN showed strongly rectifying effect with both electrode materials. All experiments were conducted on films of around 1 µm thickness and with a second gold or aluminum electrode.

Du and Chen (1998b) applied differently doped semiconducting silicon electrodes to PZT while maintaining a Pt ground electrode throughout their

sen et al. (1997); Han and Lee (2000); Cheng et al. (2000); $YBa_2Cu_3O_7$ Ramesh et al. (1992); Lee et al. (1993); Koo et al. (1999); IrO_x, Hase et al. (1998); Yi et al. (2001); $BaRuO_3$, Koo et al. (1999); $LaNiO_3$, Kim et al. (2000a); Wang et al. (2001); Chao and Wu (2001); Meng et al. (2001); Kim and Lee (2002); $SrRuO_3$, Tsukada et al. (2000)

experiments (Du et al., 1998). The fatigue effect is strongly reduced for p-doped Si and slightly reduced for n-type Si with respect to Pt electrodes on either side. A similar effect is obtained for a thin insulating layer of SiO_2, if the voltage is kept beneath a certain breakdown level. The authors ascribe the effect to a strongly reduced number of free electrons in PZT due to the different Schottky barriers. The agglomeration of oxygen vacancies is suppressed, because the mobility of purely $V_O^{\bullet\bullet}$ is assumed much lower than that of $V_O^{\bullet}$, for which the injected electrons provide the change in charge state. It has to be observed, that for Si-electrodes, the amount of charge carriers available at the PZT interface is much lower resulting in a relaxation of the whole layer structure at about 2.5 to 10 kHz. Above this frequency, the layers can no longer be switched in a single cycle.

1.11 Crystal- and Microstructure

The influence of crystal and microstructure has so far been mainly addressed by the group from the Pennsylvania State University. The most comprehensive study on fatigue in ceramic ferroelectrics prior to our work was the Ph.D.-thesis by Qiyue Jiang published in a series of articles (Jiang et al., 1992; Jiang and Cross, 1993; Jiang et al., 1994b,a,c,d,e). An important observation was that different crystal structures in hot pressed compositions $La_xPb_{1-x}(Zr_{0.65}Ti_{0.35})O_3$ (x/65/35, 5-7 μm grain size, sputtered Au) yielded different fatigue behavior (Jiang et al., 1994d). While the entirely rhombohedral composition 7/65/35 did not fatigue at all (different from the thin film results just mentioned (Lin et al., 1997), p. 31), the tetragonal compositions 8/65/35 and 8.5/65/35 fatigued rapidly. They assigned the fatigue behavior of these compositions to space charges fixing the domains. In the purely electrostrictive composition 9.5/65/35 fatigue was due to microcracking like shown in Sect. 3.7.1. It was observed that compositions exhibiting partial double hysteresis, like at the transition from ferroelectric to antiferroelectric exhibited the strongest fatigue of all. The temperature dependent data on the ferroelectric composition 7/65/35 showed the highest fatigue for the most open hysteresis loop. This particular material may exhibit a different influence of temperature on fatigue, because the hysteresis becomes significantly slimmer for temperatures approaching the Curie-point. In the electrostrictive composition 9.5/65/35 the fatigue was actually strongest at room temperature, where hardly any hysteresis is observed, while the material did not fatigue at -140°C, where is exhibits a fairly open hysteresis loop. Fatigue in 9.5/65/35 is entirely due to microcracking. As the material can compensate local stresses by ferroelastic switching once ferroelectric domains develop, the microcracking ceases. Fatigue due to point defects will not occur at low temperatures, because defect mobility is negligible.

In a third work Jiang and Cross (1993) claimed that the difference in fatigue between two batches of PLZT 7/65/35, one sintered and one hot-pressed, was due to lower porosity of the hot-pressed material. They did not discuss the additional influence of grain size on fatigue, which was considerably larger in the hot-pressed material. They explained their data by a higher amount of space charge possible in higher porosity materials. This can equally well be due to the distinctly higher grain size. The influence of grain size on fatigue was directly related to the existence of a spontaneous microcracking criterion above a certain grain size by the same authors (Jiang et al., 1994c). The fatigue was uniquely assigned to this effect.

Recent data on highly oriented films from the same group will be discussed on page 162.

Another effect was observed by Pan et al. (1989) at the morphotropic phase boundary between ferroelectric and antiferroelectric $(Pb_{0.97}La_{0.02})(Ti, Zr, Sn)O_3$. Slightly more fatigue occurs for compositions close to the phase boundary, but their assignment was vague and could be the effect of other experimental effects not considered.

An increased degree of texture or possibly also the phase transformation of some grains in ceramics during fatigue was observed by Dausch (1997). The intensity of (100) and (110) peaks in PZT-5H increased, while the (111) peak decreased during bipolar cycling (50 Hz, 1 kV/mm). This is actually far beneath the coercive field of hard materials for single cycles. It indicates a higher orientation of the domains out of their original rhombohedral polar directions, which were parallel or at least close to the direction of the fatiguing bipolar electric field. PZT-5A shows the opposite effect (50 Hz, 1.5 kV/mm), the (111) peak intensity increases, while (110) remains constant, (100) increases, and (001) slightly drops. The fact that a partial change in crystal structure was a possible cause of the experimental results was not mentioned in the text, but will be shown to be reasonable in Sect. 5.8.3.

1.12 Temperature Dependence of Fatigue

Despite the fact that almost all studies of fatigue assign a large amount of the fatigue effect to point defects, only few studies exist on the temperature dependence of fatigue.

A lower fatigue rate for higher temperatures was reported by Kudzin et al. (1975) for $BaTiO_3$. While at 25°C the fatigue down to less than 10% of the initial switching polarization already occurred within 10^6 cycles, the switchable polarization at 80°C was still higher than 65% after $5 \cdot 10^6$ cycles. Duiker et al. (1990) showed similar data on sputtered PZT 54/46 thin films (see discussion of their data on p. 38).

A detailed description of fatigue in PZT thin films was also given by Mihara et al. (1994b). They only observed differences at low cycle numbers and the identical logarithmic range for temperatures reaching from 27°C to 150°C. Like most other studies they observe the consecutive slow fatigue, logarithmic fatigue and saturation range in the remnant polarization. A higher fatigue rate for higher cycling fields was observed. The logarithm of the cycle number at 50% polarization loss scaled with the inverse of the cycling field. The difference in saturation and remnant polarization, the so called *non-switching* polarization, started decreasing in the middle of the logarithmic regime.

A detailed experimental study of thermally activated processes in fatigue was performed by Paton et al. (1997) on thin film $Pb(Zr_{0.60}Ti_{0.40})O_3$ in the temperature range from 100 K to 500 K. At 100 K no fatigue is observed up to 10^7 cycles, while the same material degrades to 0.6 of its initial polarization within 10^5 cycles at 500 K, the other temperatures leading to intermediate values. The logarithmic decay rates are unfortunately not given. An activation energy of 0.05 eV is determined. The authors tentatively ascribe this low value to an electronic process involving electrons and holes as charge carriers instead of ionic defects, for which the activation energies are around 1 eV. The work also includes the work contribution to the activation energy, which is given by the external electric field. The effective ionic conduction Δj_{net} is proportional to the degradation rate $R(N)$ with N the number of cycles. The electrical term then reads:

$$\Delta j_{net} \propto R(T) = \frac{C}{kT} \exp\left(\frac{-\Delta G + \frac{zqb\Delta E_L}{2}}{kT} \right) , \tag{1.31}$$

an expression, which will reappear in similar form in Sects. 5.8.4 and 5.8.5. Similar results had been reported earlier by Brennan et al. (1994) who experimentally determined a similar activation energy for fatigue (0.43 eV, PZT (20/80), 300 nm).

As can be seen from these three examples the change of fatigue rate not only depends on the increase or decrease of temperature, but also on the range of temperatures, which is considered.

1.13 Switching, Relaxation, and Rate Dependencies

Along with the assignment of possible microscopic origins of the fatigue phenomenon, the influence of these defects on the actual switching mechanism has been a point of discussion. In thin films two types of major fatigue mechanisms have been discussed. The first one concerns the blocking of domain wall motion, while the other one ascribes the fatigue to inhibiting the nucleation of new domain walls at the electrodes.

Colla et al. (1998b) used a particular set of cycling conditions in an attempt to discern these two scenarios. Their material fatigued after about $5 \cdot 10^7$ bipolar rectangular cycles at 30 kHz to a low fraction of the initial remnant polarization, like most other materials do, if platinum electrodes are used. The second cycling used a very slow procedure at 1.7 mHz. The cycles themselves were not rectangular but consisted of two voltage levels for each polarity. First, the external field was set to the coercive field for 300 seconds and then briefly increased to the maximum field value, where the exact time period was not specified. This double step voltage was applied in both field directions. This led to a reduction of the remnant polarization after 10 to 15 cycles. Completely bipolar switching on the other hand did not alter the maximum reachable polarization. They argued this to be a proof of the fatigue of the domain wall movement at this low frequency and excluded the interface layer inhibiting domain wall initiation to be a fatigue cause at the low frequencies. As a further argument they measured the dielectric constant at different bias fields. The pronounced maximum of the dielectric constant ϵ at the coercive field E_c in an uncycled sample was rapidly reduced, while ϵ increased at a very high bias field. An interpretation of their results in the frame of different relaxation times will be given in the discussion of the different fatigue mechanisms in Chap. 5.

A work that was very appealing on first sight was the study by Zhang et al. (2001a). They showed that the fatigue actually ceases, when the cycling field is applied at a high frequency. Unfortunately, it was not well specified, whether their samples stayed beneath the Curie temperature. As we know, our samples of 1 mm thickness, like theirs, heat up 90-100°C at 50 Hz in cooling silicon oil (see Sect. 2.1). Very effective cooling has to be provided to maintain the samples beneath the Curie point at higher fatiguing rates. It moreover seems that the ferroelectric transition temperature was exceeded in their experiment and the dielectric phase was cycled with much less fatiguing effect. As was mentioned above, also the total switching will no longer occur at higher frequencies ($>$ 10 kHz) eVen in thin films (Du and Chen (1998b); Du et al. (1998), p. 31).

1.14 Microcracking

The influence of microcracking is definitely fatal when they occur in larger amounts. Fortunately, ferroelectrics have the possibility to accommodate local stresses due to the ferroelastic switching and suppress microcrack formation to a large extent. Nevertheless, microcracking has been found in many cases and was sometimes eVen found to be dominating, e.g. in electrostrictive compositions (see Sect. 3.7.1).

In one of the early studies Salaneck (1972) showed that microcracking occurs predominantly in those rhombohedral compositions that exhibit the highest piezoelectric effect. Microcracking is stronger in coarse grained materials, because the piezoelectric strain is higher. On the other hand Carl (1975) showed that microcracking also occurs in the purely tetragonal La-doped or Mn-doped lead titanate in a 15 μm thick layer beneath the electrodes. Almost no 90° domain switching is observed in these compositions, but much 180° switching showing the high degree of clamping in these strongly tetragonally distorted grains. Spontaneous cracking occurs above a certain grain size. This is true for ceramic materials exhibiting highly distorted unit cells like $Pb(Mn_{1/3}Nb_{2/3})_{0.44}Ti_{0.44}Zr_{0.12}O_3$ ($d \geq 10\mu m$, Nejezchleb et al. (1988)), PZT ($d \geq 10\mu m$, Kroupa et al. (1989)) and PLZT ($d \geq 7\mu m$, Jiang et al. (1994c)). Dausch (1997) showed that microcracking contributes to the fatigue in PZT-5A, but not in PZT-5H, which was eVidenced by electromechanical resonance spectra. As these data were partially obtained in RAINBOW actuators[4], it is not clear whether some of the effect is due to the gradient in point defect concentration and the resulting stress gradients.

1.15 Models

There has been about a decade of intense discussion about the fatigue mechanism in ferroelectrics. A brief summary of possible mechanisms in the context of thin films was given in the book by Scott (2000) and the article by Tagantsev et al. (2001).

An early and elegant model for fatigue as a bulk phenomenon was developed by Brennan (1993). The model uses the standard description of a uniaxial ferroelectric in Landau-Devonshire theory (Devonshire, 1954). The incorporation of a point defect would result in an increase of free energy, if the domain were to maintain its original polarization direction. As a consequence a reorientation of the polarization occurs directly at the defect. In a uniaxial model the generation of a head to head or tail to tail 180° domain wall is a necessary consequence with the point defects located at the wall. The formal reduction in free energy is achieved by assigning a minimum area of domain wall to the point defect for which the wall energy is reduced. As a second result the logarithmic development of fatigue is obtained in the form

$$P - P_o = \frac{kt}{c} \log(N) \tag{1.32}$$

with N the cycle number. The assumption is that, at first, point defects of the same kind are highly attracted to the vicinity of the already existing defect, because the head to head / tail to tail domain wall extends beyond the defect itself, equivalent to a net charge of opposite sign in a ring around the defect. At this location additional defects accumulate with a certain activation energy. For a rising size of the defect agglomerate, the energy barrier increases for each new defect to be incorporated. The activation energy increases and the accumulation becomes less and less probable yielding less and less agglomeration with time for an existing defect. In turn a certain amount of domain wall freezes along with the defect and grows with time

[4] RAINBOW actuators are disc shaped actuators, mostly of PZT, which were chemically reduced on one side to significantly change the piezoelectric effect across the thickness. The strain is obtained along the thickness direction of the plate due to the warping of the plate (Haertling, 1994).

accordingly. The frequently observed logarithmic decay in remnant polarization with cycle number results as well as a reduced ferroelectric hysteresis. The Landau-Devonshire theory yields the hysteresis form

$$E(P) \approx \alpha P + \beta P^3 \tag{1.33}$$

$$E_{tot}(P) = E(P - \Delta P_{vacancies}) + E(P + \Delta P_{vacancies}) \tag{1.34}$$

in which the total polarization is modified according to the frozen polarization due to vacancies. This formulation yields a material description, for which the remnant polarization decreases like it is observed in actual fatigue experiments leading to a linear dielectric in the extreme case, but the coercive field drops along with this development, which is not the case in experiments.

The model by Brennan was extended by Shur et al. (2000, 2001b,c). Their agglomeration model is based on the Kolmogorov-Avrami theory for phase changes (Kolmogorov (1937); Avrami (1939, 1940, 1941); Ishibashi and Takagi (1971)). Primarily, seeds of a new phase are formed with a certain probability. Then these seeds start growing. The model is based on a 2D rectangular net of cells permitting two opposite signs of polarization. For the boundary propagation between up and down orientation, two types of wall motion are permitted, linear propagation and edge growth. Without any freezing of certain domains, this model describes the switching current for polarization inversion in ferroelectrics in general. The local field is determined by the external field and the local field. This local field is modified by each switching cycle. The growth mechanism becomes fractal in character. Initially, the primary seeds extend all over the two dimensional plane. Each seed then starts growing. Considering a rectangular voltage change for switching, the switching current for a given nucleation and growth probability exhibits a maximum distinctly after the discrete voltage step. After a certain number of cycling the local fields have sufficiently increased to shield the external field beneath the locally necessary coercive field. These domains become entirely blocked. This then reduces the area of possible nucleation sites of the opposite polarization. Domain growth still occurs, but predominantly in one direction. The switching current looses its maximum and becomes smeared out to longer times. During the fatigue process the fractal character of the nucleation and growth process changes from a 2D phenomenon to a 1D domain wall propagation scenario. The switchable polarization in total decreases all over. The crucial parameter introduced to describe fatigue is again a fraction of non-switchable domain region. This also tends to grow with a certain probability. Finally the model leads to such a limited number of nucleation sites, that switching will only occur at the rim of the frozen domains. The total switching is severely hampered (see also Sect. 5.8.4).

One of the first quantitative models for fatigue where the point defect motion was directly incorporated is due to Yoo and Desu (1992c) (see also Sect. 5.8.4, p. 146). They used the point defect flux densities to calculate the net point defect flow towards one of the electrodes. Two assumptions were introduced. According to experiment the major blocking of defect motion was considered to occur at the electrodes, because the model was designed for thin films only containing few grains. The second assumption was that local fields will immediately alter the net flux densities between both field directions. Using standard rate equations (Hench and West, 1990) they arrive at the logarithmic decay of polarization and obtain an average jump distance for the point defects of about one lattice parameter, which rendered their model very sensible. The application of RuO_x electrodes supported their model, because the trapped oxygen vacancies can be liberated into the electrode. The data by Du and Chen (1998b) were not known at that point, where merely the appropriate majority carrier in the electrodes determines whether strong fatigue occurs or not.

An agglomeration model not relying on the exchange of point defects with the environment was presented by Dawber and Scott (2000a). They utilized the already mentioned assumptions (Yoo and Desu, 1992c; Brennan, 1993) that point defects agglomerate in planar structures. For a tetragonal crystal containing a local polarization vector along the long axis of the unit cell the six next neighboring positions for a defect ion are not equivalent as was first developed by the works of Arlt et al. (Robels et al. (1995b); Robels and Arlt (1993); Lohkämper et al. (1990); Arlt and Neumann (1988)). Thus the defect flux density is not equivalent along the different crystal directions and is particularly different between the positive and negative polarization directions. The model further assumes a thin layer of accumulated oxygen underneath the electrodes. This layer then generates a screening field reducing the local field in the interior of the thin film. The model does not differentiate between the two electrodes, thus a net drift will occur towards both electrodes and the driving force is given through the modified gradient in defect concentration on top of the external electric field. The general dependence of fatigue in thin films on temperature (Paton et al. (1997), see p. 34), cycling voltage (Mihara et al. (1994a)) and frequency (Colla et al. (1998b)) are reproduced by the model.

Two aspects of fatigue in ferroelectric thin films were modelled by Duiker et al. (1990). The first aspect concerned the formation of dendritic trees of semiconducting defects growing into the depth of the ferroelectric films. They used a Monte Carlo method to simulate the dendrite growth. The model predicted the observed dependence of polarization on cycle number via the short circuiting of grains in the open interior of such dendrite trees no longer contributing to the ferroelectric effect, and the final breakdown of films at high cycle numbers. EVen though the model yields reasonable curves for the polarization, this approach has not been reused in later models of fatigue.

A second aspect in their work were the switching kinetics of the domain system at different fatigue stages. They adopted the nucleation and growth model for domain switching developed for infinite grains by Ishibashi and Takagi (1971); Ishibashi (1985); Duiker and Beale (1990) including the modified version describing ferroelectrics of finite size and modelled the switching of KNO_3 thin films. The major outcome from fits of the model to experimental data is a change of character of the switching dimensionality depending on film thickness, but they do not correlate this to fatigue data like Shur et al. (2000) do.

In a recent work Scott and Dawber (2000b) suggest the agglomeration of $V_O^{\bullet\bullet}$ into linear chains, which then grow to planar defects. They propose a ^{57}Fe Mößbauer experiment with highly diluted ^{57}Fe which becomes localized at the ends of linear chains of oxygen vacancies. For the fatigue the Mößbauer signal should vanish during the growth of the planar defects. This idea and their above mentioned model assumption on defect agglomeration (p. 38) were crucial stimuli for the model developed in Sect. 5.8.4.

The different models will be thoroughly discussed in view of the results from this study in Chap. 5.

1.16 Fatigue-Free Systems

The first work to show a fatigue free system in thin film ferroelectrics was the work by Desu and Vijay (1995). They showed that pulsed laser deposited oriented thin films of $SrBi_2(Ta_xNb_{2-x})O_9$ (SBTN) do not fatigue. The switchable polarization of these films is not oriented along the crystal axes as is usually the case for the symmetry of orthorhombic crystals, but slightly tilted. An explanation for the lack of fatigue was not given and is still not clear (see extended discussion in Sect. 6.10).

Despite the usage of Pt-electrodes yttrium-doped PZT showed no fatigue of the remnant polarization up to 10^{10} bipolar square cycles at 1 MHz. This was ascribed to the donor character of Y substituting Pb in the perovskite type structure (Kim and Park, 1995). But this is similarly true for all other donor dopants, like lanthanum for which fatigue was readily observed for Pt electrodes (Kim and Park, 1995). Within the people working on fatigue, it seems not to have been noticed, that Y^{3+} as well as most of the rare earth ions, except for the very large La^{3+} and Pr^{3+}, can each be incorporated at both lattice sites, lead and titanium. The very recent modeling analysis by Buscaglia et al. (2001) (see also the many references therein) explicitly showed that Y^{3+}, Er^{3+}, Tb^{3+}, and Gd^{3+} will behave so in $BaTiO_3$ and that the ratio at which they substitute the host ions depends on the Ba/Ti-ratio in order to yield a fully compensated structure. This was experimentally proven in $BaTiO_3$ by Zhi et al. (1999), who found that up to 1.5% of Y will be incorporated at Ba-sites while up to 12.5% at titanium sites. For concentrations higher than 6% the tetragonal distortion vanishes. Thus, incorporation of Y-concentrations up to about 1% are suiTable to compensate any off-balance in charge state of the ceramic. Scott et al. (1991) explicitly showed a drastically reduced conductivity of yttrium doped PZT thin films. The conductivity was three orders of magnitude lower than in an undoped material. Despite the advantages of yttrium as a dopant, the restricted maximum solubility at the Pb-site does not allow an absolute protection against fatigue. EVen in Y-doped PZT the coercive field rises during bipolar switching (Kim and Park, 1995) and thus some form of microscopic fatigue does occur eVen in this fully compensated material.

2 Macroscopic Phenomenology

The introductory chapters have already revealed that the fatigue phenomenon is a complicating interplay of point defects, microstructure, domains, and the external boundary conditions. This chapter is devoted to a summary of macroscopic data on one particular material during electrical fatigue, namely PIC 151. Many of the known results are reproduced in this material, but a series of measurements so far not described in literature round up the set of macroscopic phenomena encountered around fatigue. The corresponding microscopic observations are then displayed in Chap. 3.

2.1 Fatigue and Measurement Procedures

Cycling:

The material was fatigued using an electric transformer $0 - 220\,\mathrm{V}$ of line voltage at 50 Hz. The output was transformed by a second transformer of fixed conversion factor 10, permitting a range of voltages from 0 to 2200 V. The samples were fixed by two copper clamps and immersed in silicon oil. For the fatiguing experiments none of the two copper clamps was grounded in order to avoid any possibility of an experimentally induced bias. The temperature of the samples was monitored in a selected set of measurements using a thermocouple intensely bonded to the sample by heat conducting glue at the ground side. For these particular measurements one of the copper clamps was grounded. The samples heated up to 80°C during cycling.

For unipolar cycling the setup was extended by a bridge rectifier. The samples were thus cycled at 100 Hz unipolar with a $|\sin(\omega)|$ waveform. As the bridge rectifier has very high impedance in blocking direction, a suitable resistance had to be added parallel to the sample for permitting the voltage to actually drop back to zero.

A third setup was used for mixed loading fatigue (Fig. 2.1). A uniaxial mechanical compressive load was provided at 25 Hz sinusoidal by a hydraulic testing machine (Materials Test Systems 810.22, MTS, Materials Test Systems Corporation, Minneapolis, Mn, USA). The load was monitored using a 5 kN load cell. To ensure uniaxial compressive stresses two metal (aluminum, brass, or steel) cylinders of 10 mm diameter and 8 mm height were introduced

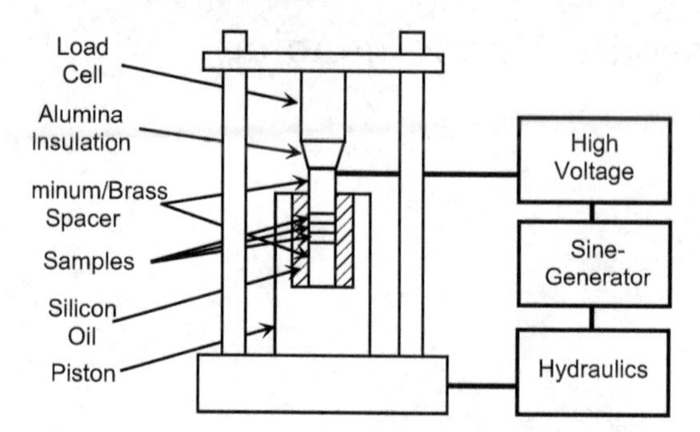

Fig. 2.1. Setup for mixed mechanical and electrical fatigue loading

one on either side of the sample. The elastic constants of these blocks differed only slightly (aluminum, brass) or much (steel) from those of the PZT samples with the aluminum slightly softer and the brass slightly harder. If a soft material is pressed against a much stiffer piece of the device, a high component of hydrostatic stress is induced. The hydrostatic component of stress induced by the stiff hydraulic piston is confined to a cone of 90° tip angle and has thus completely vanished at the location of the sample, if aluminum or brass spacers are used (at least this was the goal, see Sect. 2.7). The high voltage was provided by a high power high voltage bipolar amplifier (HCB 500M - 10000, F.u.G. Elektronik GmbH, Rosenheim, Germany). It was synchronized to the load by a phase shifter of a lock-in amplifier (SRS 830DSP, Stanford Research Systems, Stanford, California). The unipolar electric field was highest for zero stress and zero for maximum compressive stress.

Measurement:

The electric polarization was measured using a calibrated (4 digits) 4.7 µF capacitor in series with the sample. As the sample had a total capacitance of about 1 nF, the error in voltage was about $2 \cdot 10^{-4}$ and neglected in the measurements. The voltage reading on an electrometer (6517, Keithley Instruments, Cleveland, OH, USA) provided the total charge flow into the 4.7 µF capacitor and is thus a direct measure of the polarization of the sample in series. The internal resistance of the electrometer is 10^{14} Ω.

A linear variable displacement transducer provided the strain measurement (Hottinger Baldwin Messtechnik, Darmstadt, Germany). Its resolution is 20 nm. The largest error in strain measurement arises due to a possible warping of the sample. Certain displacement measurements were performed between two sharp alumina tips each covered with silver paint after mounting the sample to provide electrical contact. These latter measurements assured that the strain irregularities, which will be shown in the following sections, are not experimental artifacts. Most displacement measurements were conducted directly between the surface of the microphone and a single alumina

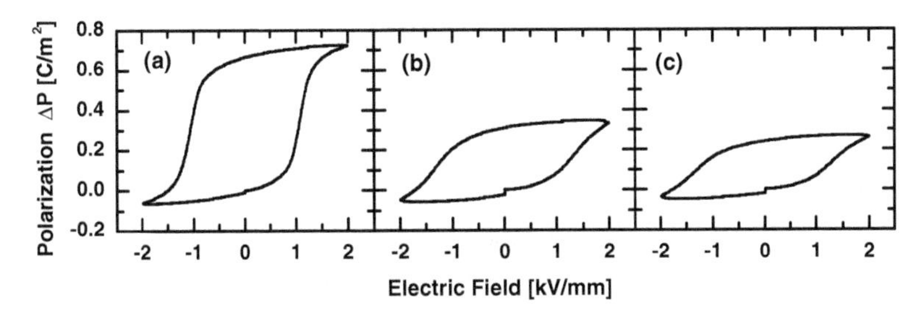

Fig. 2.2. Loss of polarization due to bipolar cycling (1.96 kV/mm, 50 Hz). Fields of ±2 kV/mm were applied during measurement. (**a**) 0 cycles, (**b**) $3 \cdot 10^6$ cycles, and (**c**) 10^8 cycles

tip to be able to monitor the acoustic emissions simultaneously (see Sect. 4.1.2 and Fig. 4.1). Only certain measurements were confirmed using the double tip setup. The actual surface of the microphone was covered with a thin aluminum foil to ensure electrical contact. Despite the possible formation of an insulating alumina layer on the aluminum foil it was used for a better acoustic coupling, because it is much softer than for example brass and allowed a close acoustic contact to the microphone by simply pressing it by hand.

All measurements of strain and polarization were performed at 20 mHz unless otherwise stated to also ensure good quality of the acoustic emission data.

2.2 Polarization and Strain Loss

Like in most other PZT compositions of either bulk or thin film geometry fatigue in PIC 151 is first observable in a loss of the height of the polarization hysteresis as shown in Fig. 2.2 and a more subtle increase in the coercive field (see references in the introduction, p. 18). Severe fatigue due to bipolar cycling occurs, if the cycling field exceeds the coercive field. Figure 2.3 shows the loss of the remnant polarization with cycle number for different cycling fields. The first important thing to note about this behavior is the correlation of fatigue with the cycling field values. Only fields exceeding the coercive field yield considerable fatigue. The microscopic fatigue mechanism thus has to be triggered by high stress or field states in the microstructure, which occur, once the domain system is no longer available to accommodate the externally applied fields.

The loss in strain appears even more drastic (Fig. 2.4). Irrespective of the first poling direction one of the wings of the strain hysteresis S_d (degrading) degenerates more rapidly with increasing cycle number than the other one S_wd (weakly degrading). Once the material degrades stronger on S_d, this part of the hysteresis remains the more damaged one in all following cycling steps.

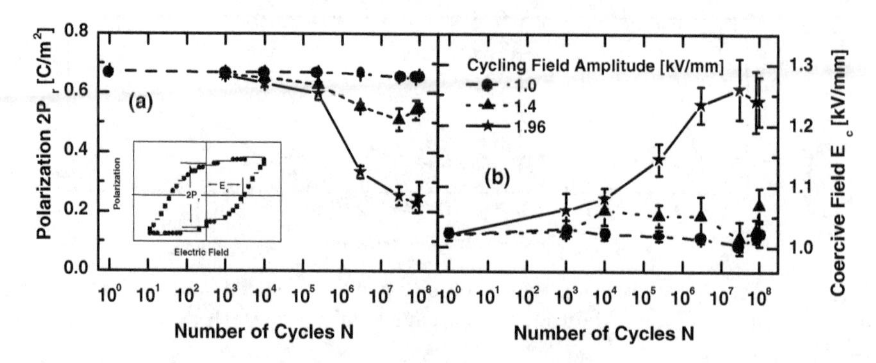

Fig. 2.3. Development of remnant polarization (**a**) and coercive field (**b**), for different cycle numbers and at different maximum cycling field

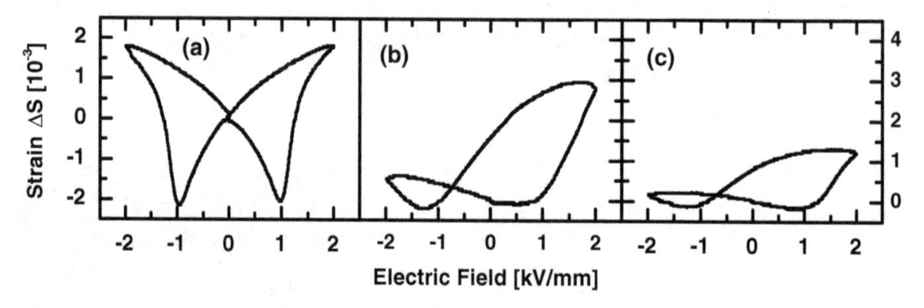

Fig. 2.4. Asymmetric loss of strain in PIC 151 due to bipolar cycling (1.96 kV/mm, 50 Hz). Fields of ± 2 kV/mm were applied during measurement. (**a**) 0 cycles, (**b**) $3 \cdot 10^6$ cycles, and (**c**) 10^8 cycles

Cycling at 1.96 kV/mm yields the strongest fatigue, while for 1.0 kV/mm almost no fatigue is observed. Cycling at 1.4 kV/mm generates values in between. In Fig. 2.3 the development of the coercive field (a) and remnant polarization (b) are shown. E_c steadily increases with cycle number from $E_c=$ 1.02 kV/mm at 0 cycles up to 1.26 kV/mm at $3 \cdot 10^7$ cycles. The degradation for P_r occurs slightly later than first changes in the coercive field and the major reduction occurs between $2.5 \cdot 10^5$ and $3 \cdot 10^6$ cycles down to 0.33 C/m^2. The strain values of the strongly decreasing strain S_{d} start degrading at approximately the same cycle numbers as P_r. Similar logarithmic fatigue up to $3 \cdot 10^6$ cycles and subsequent decay even including a small recovery at 10^8 cycles occur. The strain for the less degrading strain S_{wd} only starts decreasing above $N > 3 \cdot 10^6$ (E_{cycle}=1.96 kV/mm). Cycling at 1.0 kV/mm again leads to no decrease throughout the whole range of cycle numbers. For cycling at 1.4 kV/mm, S_{wd} and S_{d} can not effectively be discerned. Both branches degrade from $3.9 \cdot 10^{-3}$ at 0 cycles to $2.4 \cdot 10^{-3}$ at 10^8 cycles. The

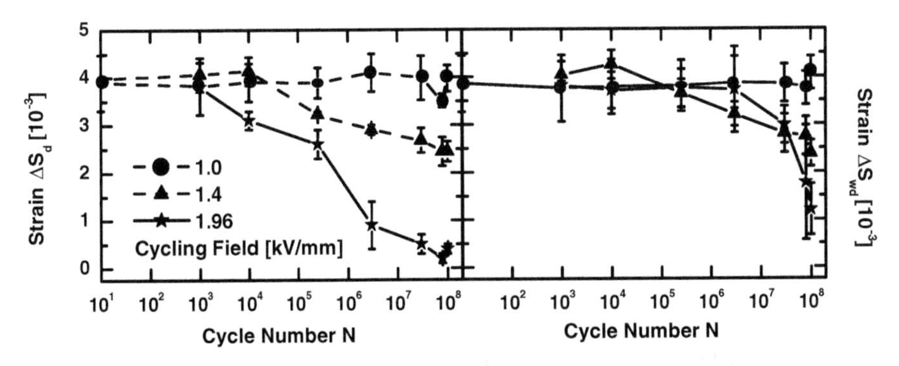

Fig. 2.5. Fatigue cycle number dependent maximum strain values for the strongly decreasing (ΔS_{d}) wing and the weakly decreasing wing (ΔS_{wd})

asymmetric degradation thus seems to be a feature pertinent to higher cycling fields.

The same sequence of fatigue stages is found in PIC 151 as in other bulk and thin film soft PZT compositions, an incubation period up to $2.5 \cdot 10^5$ cycles, a logarithmic fatigue range between $2.5 \cdot 10^5$ and $3 \cdot 10^6$ cycles and a range of saturation above 10^7 cycles (Yoo and Desu, 1992c; Pawlaczyk et al., 1995). PIC 151 thus behaves like a regular PZT material and the results in this work seem likely to be representative of a large number of compositions at the morphotropic phase boundary.

Some implications on the microscopic mechanisms are already evident from these results.

Firstly, the reduction of strain is correlated to the reduction of polarization. Thus a major part of the fatigue mechanism has to deal with reducing the mobility of 90° domain walls.

Secondly, fatigue was found to strongly depend on cycling voltage. All macroscopic parameters showed less fatigue at lower fields and field levels around E_c did hardly any harm. For about 1.4 E_c the logarithmic fatigue regime does not occur as evidently as for 2 E_c cycling voltage. Whatever microscopic mechanism is present, it is essentially induced when the material is driven far into saturation. This implies that high internal stresses and fields aid inducing the types of defects, that will hinder the domain wall motion.

At this point already a brief comparison to thin film results is given. A first significant difference is that in thin films fatigue was also observed to occur beneath E_c (White, 2000). The definition of P_r and E_c in thin films is more difficult than in bulk materials, because the hysteresis curve is generally more slanted, and P_r only reaches about 30% of the saturation value $P_{\mathrm{sat,unfatigued}}$ at high fields (Warren et al., 1995b), which is due to the high degree of mechanical clamping by the substrate (Khatchaturyan, 1995). During fatigue the slant angle significantly changes while E_c remains essentially unchanged.

In bulk materials P_r reaches 80% of P_{sat} and the electrical hysteresis remains fairly squared. Furthermore, Bobnar et al. (1999) observed the symmetric reduction of the strain hysteresis in PLZT 8/65/35 thick films (10^6 cycles, Au-electrode). Considering these strong differences, it is surprising, that the numbers of cycles to generate a similar decrease in P_r are fairly comparable for thin films with metal electrodes and bulk materials. Thus, some similar microscopic mechanism may occur in both bulk and thin films and determine the general fatigue to some extent.

2.3 Asymmetry and Offset-Polarization

2.3.1 Strain Asymmetry

An asymmetric degradation of the strain hysteresis like the one of Fig. 2.4 has previously been observed by Weitzing et al. (1999) and interpreted as a remaining influence of the first poling in prepoled samples. Their result is identical to the one shown here, but the experimental procedure was different. Thermal treatments before and in between the cycling steps were introduced in our measurement (4 h, 400°C) to remove a preferred orientation due to the first poling. The side of stronger degradation S_{d} actually differs with respect to the first poling direction for different samples. Nevertheless, one of the branches S_{d} degrades more rapidly than the other one and remains degraded more strongly despite thermal treatment. In an earlier measurement an asymmetry of displacements was observed for a two layer model geometry modeling a multilayer geometry of cofired stack actuators (Furuta and Uchino, 1993). There the asymmetry was undoubtedly due to the formation of cracks on one side of the sample and also interpreted that way.

A simple macroscopic explanation for our data is the development of an offset-polarization π (Nuffer et al., 2000). According to Landau-Devonshire theory, the strain of an electrostrictive material is proportional to the square of the total polarization (Grindlay, 1970). Even though this is strictly true for single domain systems near the paraelectric-ferroelectric phase transition (Lines and Glass, 1977; Xu, 1991; Grindlay, 1970), this material description still holds amazingly well far away from the phase transition temperature and as a macroscopic approximation of ceramic behavior. In this case we have to consider all contributions to polarization, thus the switchable and the offset polarization (also termed imprint polarization in thin films). This leads to:

$$S(E) = Q(P(E) + \pi)^2 \,. \tag{2.1}$$

An approximation of the asymmetric butterfly curve from Fig. 2.4 is given in Fig. 2.6 (b) using the parameters $Q = 3.4 \cdot 10^2$ m^4/C^2, $\pi = 0.07$ C/m^2. The degree of asymmetry depends on the value chosen for π. A calculation assigning one volume fraction, (v), to a frozen polarization only exhibiting the

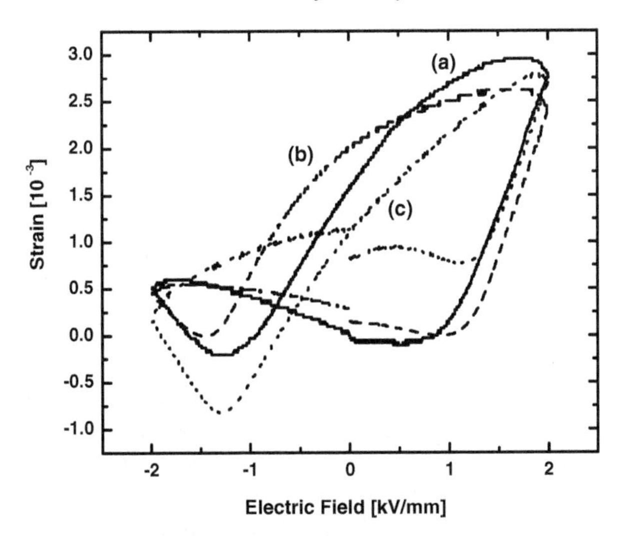

Fig. 2.6. Fit of the offset-polarization to strain. (**a**) measured strain, (**b**) fit according to (2.1), (**c**) fit considering two volume fractions (see text)

piezoelectric effect and another one, $(1-v)$, to switchable polarization showing quadratic electrostriction, $S(E) = vS_1(E) + (1-v)S_2(E)$, $S_1(E) = d_{33}E$, $S_2(E) = QP^2$, does not yield the appropriate shape of the strain hysteresis. Both quadratic wings necessarily stay symmetric only slanted by the linear term (curve (c) in Fig. 2.6). The reduction of amplitude of merely one of the wings is only explicable in the former case.

An offset polarization is distinctly different from a bias field, which is generated throughout the volume of a still switchable domain. A bias field would also be observable in the electric hysteresis, which remains strictly symmetric with respect to the electric field throughout our experiments. In some thinner samples (0.5 mm thickness) a slight bias field was observed for the first measuring cycle after fatigue *and* after the samples had been given time to age after removal from the fatiguing setup. Thus a bias field is the result of aging under no or d.c. external loading and not due to cyclic fatigue. The offset field disappeared in the subsequent cycles, while the offset-polarization was stable. Thus, a stable offset polarization develops during bipolar fatigue cycling. Direct observations previously showed fully clamped polarization in thin films using atomic force microscopy (Colla et al., 1998a) and are the basis for the fatigue models by Shur et al. (2000, 2001a,b) (see Sect. 5.8.4). Regions of fixed polarization orientation become stable and tend to orient their neighbors in only one particular polarization direction, an agglomeration process. As was pointed out by Warren et al. (1996b) local alignment of defect dipoles in the sense of the model description by Robels et al. (1995a) will also yield an offset polarization, if other local charges like electrons or holes screen the electric field generated by the dipoles somewhere remote from the dipole itself, but still within the microstructure. Warren et al. (1996b) also point out the implications of an offset polarization on strain, but no published data are referenced.

A direct measure of an offset polarization was given for thin films by Kholkin et al. (1996). Under bipolar as well as unipolar cycling their thin films (260 nm thick, PZT 45/55, Pt electrodes) fatigued within 10^7 cycles (30 kV/mm, 8.6 ms high voltage rectangular, 0.1-15 kHz). Their d.c. field dependent d_{33} values showed the ordinary hysteresis with the applied d.c. field, but superposed a polarization offset as we observed it in our macroscopic data. Thus the fixed polarization (completely pinned domains) measured by Colla et al. (1998a) using AFM directly correlates with the change in the piezoelectric coefficient. The authors furthermore observed, that the offset polarization is induced under bipolar fatigue, while unipolar fatigue induces an offset-field. Offset polarization thus obviously also occurs in thin films and is not just a bulk phenomenon. After their cycling procedure (10^{10} cycles) they furthermore observed that the remnant piezoelectric coefficient changed by 20 % within a few thousand seconds due to aging. The authors furthermore discuss, that the mere alignment of defect dipoles in the bulk of the grains is not a sufficient mechanism to account for the offset polarization, because 10^{21} cm^{-3} dipoles would be needed to generate the effect. They assign some internal domain pinning mechanism to be responsible for the effect. An asymmetry in d_{33} had previously been reported by the same group, along with data on polarization hysteresis and permittivity under d.c. offsets, which both stayed symmetric (Colla et al., 1995).

The good correlation of the cycle number in our data for logarithmic fatigue for S_{d} as well as P_r between $2.5 \cdot 10^5$ and $3 \cdot 10^6$ cycles for 1.96 kV/mm cycling field supports a common mechanism that reduces switchable polarization and strain simultaneously. We are thus microscopically concerned with a mechanism suppressing the 90° domain switching.

2.3.2 Obstacles to 90° Domain Switching

From the results in this section and Sect. 2.2 a first fairly reasonable microscopic model can be derived. Bipolar fatigue induces microscopic obstacles to the 90° domain wall motion. These obstacles grow during fatigue and progressively reduce the mobility of the domain wall system. Certain domains will entirely loose their mobility, while others will need a higher external driving force for surpassing the obstacles they encounter.

To test the assumption of such barriers to the domain wall motion constituted by some microscopic defects, two experiments were devised, one which increases the driving force on the domains, namely an increase of the electric field during measurement, and another one modifying the structure of the defects thermally (Verdier et al., 2002).

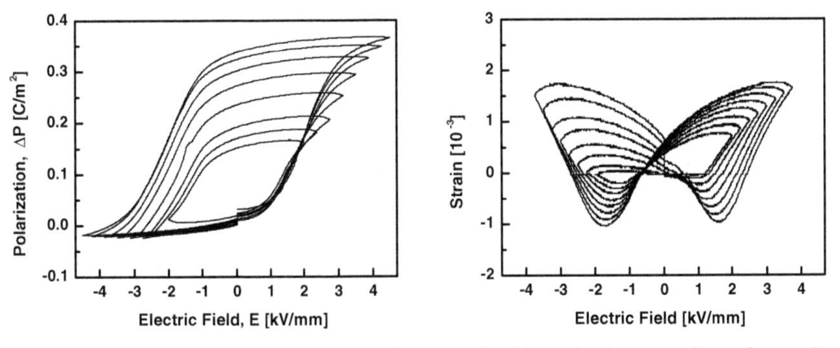

Fig. 2.7. Response of a fatigued sample of PIC 151 to fields exceeding the cycling field ($1.9\,E_c$) after $4 \cdot 10^7$ bipolar cycles. (**a**) Polarization hysteresis and (**b**) strain hysteresis

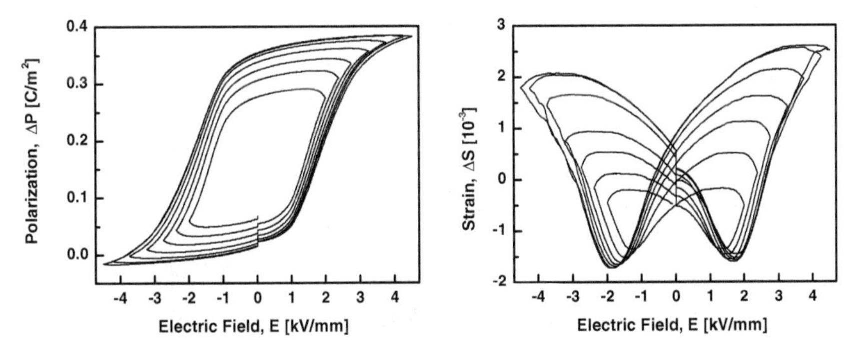

Fig. 2.8. Relaxation of a fatigued sample after the high voltage treatment of Fig. 2.7. (Left) Polarization hysteresis and (right) strain hysteresis

Fatigued samples were subjected to increasing maximum fields during a few measuring cycles. Figure 2.7 shows how polarization and strain regain almost their full amplitude at 4 kV/mm. This is similar to results by Pan et al. (1992b), who showed that high fields can recover polarization lost during fatigue. When the maximum field is reduced again, both hysteresis curves drop back to some smaller amplitude (Fig. 2.8). However, the state after this procedure is different from before, because the strain hysteresis now appears symmetric and the polarization hysteresis is centered about an effectively zero internal offset polarization.

In a second set of experiments the samples were annealed after the fatiguing procedure. Figure 2.9 displays the development of the remnant polarization measured at room temperature after annealing steps at the temperatures indicated (1 h at maximum temperature, 8°C/min heating and 3°C/min cooling). After a thermal annealing at 500°C the remnant polarization recovers 70% of its unfatigued amplitude, with no further changes at higher annealing

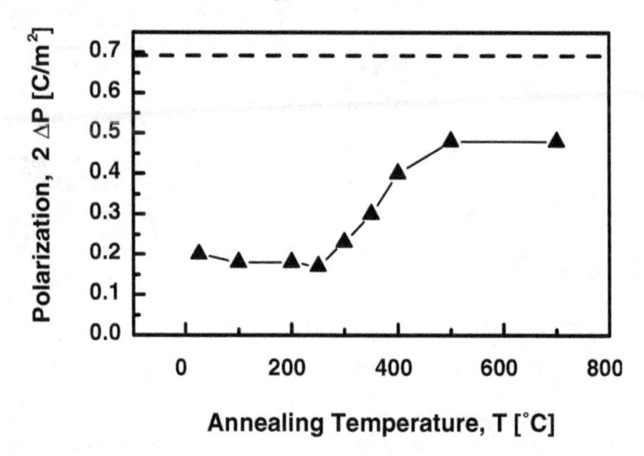

Fig. 2.9. Thermal annealing of the defects inducing the polarization drop during fatigue. The remnant polarization at room temperature after thermal annealing at the temperatures indicated

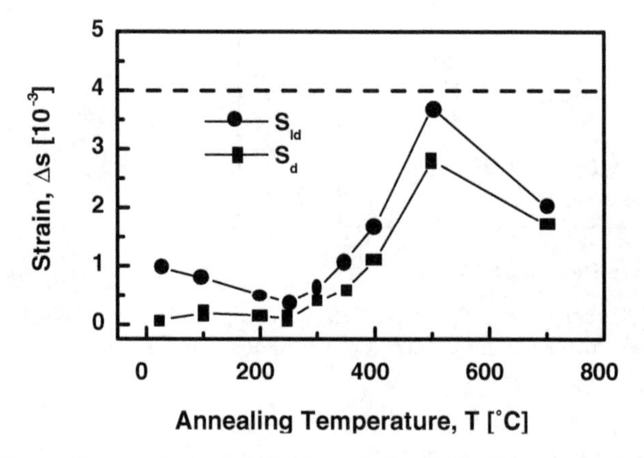

Fig. 2.10. Thermal annealing of the fatigue induced strain changes. The maximum strain at room temperature after thermal annealing at the temperatures indicated for the degrading S_d and less degrading S_wd wing

temperatures. This corresponds fairly well with the temperature assumed for freezing the mobility of oxygen vacancies in $SrTiO_3$, 700 K (Waser, 1994). At 770 K (500°C) the mobility of oxygen vacancies can thus be sufficient to disassemble the defects that developed during fatigue. The thermal annealing is suitable to remove a major part of the fatigue induced obstacles to domain wall motion. The remaining 30% of lost polarization can not be recovered thermally. Considering the results from the high field measurements, thermal annealing reaches similar values of remnant polarization as does the high field treatment. Thus, those defects which constitute significant obstacles to the domain wall motion at the cycling field, but not sufficient obstacles at higher fields are removed during thermal treatment of the sample. The major difference is, that after the high field treatment the amplitudes of the hysteresis curves drop back to values close to those immediately after fatigue. Thus, the obstacles can be surpassed by high fields, but not removed. Already Stewart and Cosentino (1970) noticed that application of elevated temperatures after fatigue would recover fatigue induced polarization losses. In a note on effects of porosity on fatigue Jiang and Cross (1993) later showed that some recovery of the fatigued hysteresis loop can be obtained in PLZT 7/65/35 at 350°C.

The remaining part of lost polarization and strain after either treatment can not be recovered in our measurements. A mechanism like microcracking, which is not recoverable at the temperatures considered, seems reasonable in this case, or mechanisms, that carry point defects over larger distances, which will not equilibrate during the annealing procedure.

2.4 Anisotropy

Pan et al. (1992c,b) were first to recognize that fatigue in bulk ferroelectrics induces a highly textured loss in switchable polarization. After fatigue of rectangular bars of ceramic PLZT (7/62.5/37.5) the polarization fatigue had dropped to about 25% of the unfatigued value. Perpendicular to the fatiguing direction the polarization hysteresis was unaltered. They then cycled in the perpendicular direction and observed that fatigue along this axis occurred earlier than in the fresh sample. An intermediate angle yielded an average of both directions. A spatial distribution of the fatigue phenomenon in thickness direction was not apparent to them for slices cut from the fatigued sample. They concluded, as we do, that the fatigue is predominantly a bulk effect in ceramic PZT's. Furthermore, they found no significant difference in material properties for either the original fatiguing electrodes or newly applied electrodes after removal of the cycling electrodes. The ferroelectric-electrode interface just only plays a secondary role in the fatigue of bulk PZT.

In Figs. 2.11 and 2.12 polarization and strain data measured in different directions of the sample are shown. In each experiment strains along two directions were measured along with the polarization and the applied electric field. Figure 2.11 shows the data of a fatigued sample with the field direction along the cycling direction (labelled Z). The severe reduction in polarization and strain amplitude is evident with respect to the undamaged sample on the right hand side. For the measurements in Fig. 2.12 the electrodes of the sample were removed, rectangular bars were cut from the disc shaped samples and electrodes perpendicular to the original cycling direction were applied. The electrical hysteresis almost reaches values of an unfatigued sample as already reported by Pan et al. (1992c,b). In our data around 80 % of the original polarization are achieved. The strain data convey the same message. Values of around 70 % of the original strain are found.

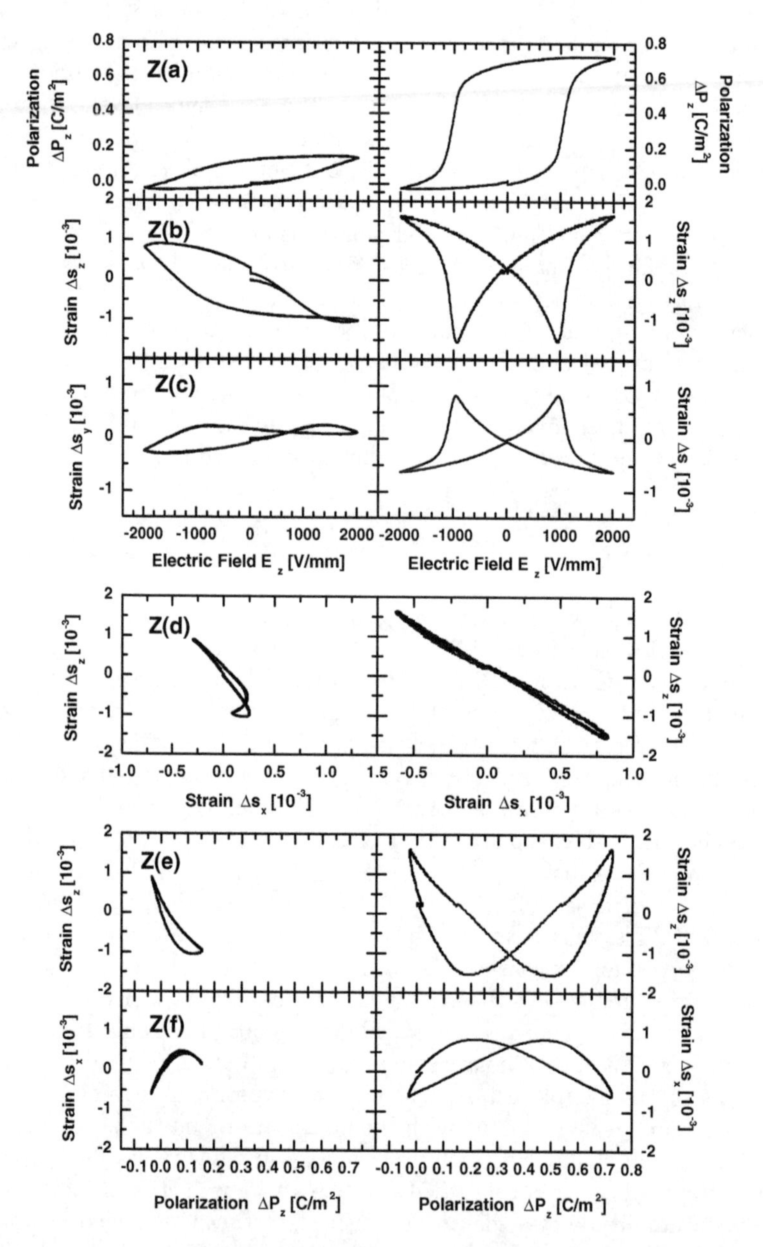

Fig. 2.11. Strain and polarization anisotropy with the external field applied *in cycling direction z* (Z(a) through Z(f)). (Left) after fatigue ($2 \cdot 10^7$ cycles), (right) before fatigue for comparison

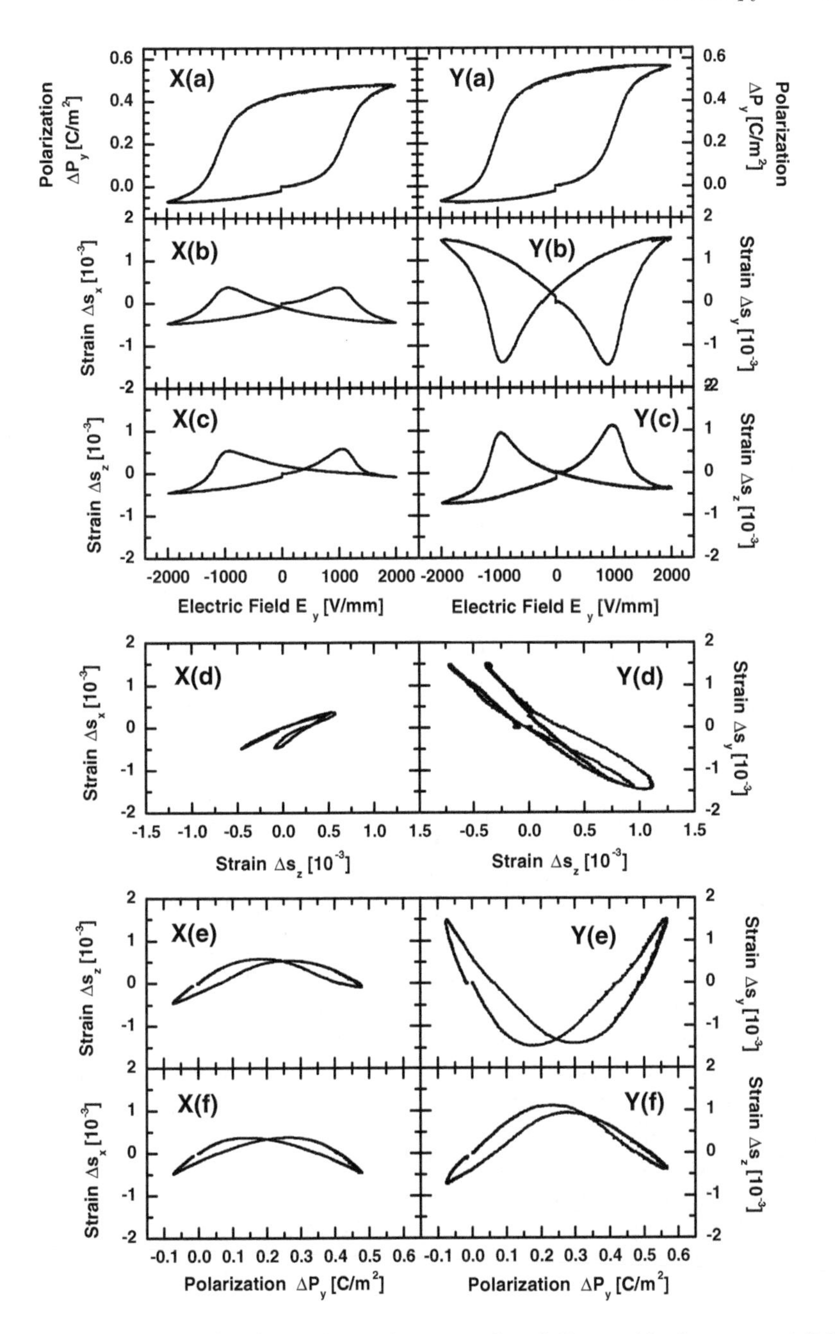

Fig. 2.12. Strain and polarization anisotropy after fatigue with the external field applied *perpendicular to the cycling direction, along y*. Measurement of both transverse strains, s_x and s_z (X(a) through X(f)) and longitudinal and transverse strain s_y and s_z (Y(a) through Y(f))

Another interesting facet of the strain data is the plot of the electrostrictive hysteresis, Fig. 2.11 Z(e) left. It is just evident that the positive polarization values are never reached, which would form the right hand part of the electrostrictive parabola, irrespective of the absolute measurement-zero for polarization, which is arbitrary due to the measurement procedure. Altogether the transverse strain still reaches about 50 % of the strain in field direction, as expected from constant volume. But it is also clear from Fig. 2.11 Z(d) left, that different from the unfatigued case, where a straight line is observed (Z(d) right), a significant hysteresis occurs.

Comparing the electrostrictive hysteresis in an unfatigued sample (Z(e) right and Z(f) right) and the measurement along the perpendicular direction (Y(e) and Y(f)), the hysteretic opening of the loops is larger in the unfatigued case. This is an indication, that a certain part of the 90° switching was already suppressed due to fatigue, despite the almost normal polarization hysteresis. It should also be marked that the slope of the polarization hysteresis (Z(a) right), or in other words the sharp tipping of the strain hysteresis in the strain data of an unfatigued sample (Z(b) right), are already smeared out in the perpendicular direction in a fatigued sample (Y(a) and Y(b)).

Irrespective of the direction of the applied field the strain along the cycling direction is asymmetric. This is particularly well visible in the plot of the transverse vs. longitudinal strain ((d) in of plots). Only the unfatigued sample shows the expected strictly linear behavior (Fig. 2.11 Z(d) right) and the lines of both polarities coincide. In all other cases an opening and a strong asymmetry between both polarities of the applied field is visible, which translates into a deviation from a straight line and the different effective slopes of both polarities (Figs. 2.11 Z(d) left and 2.12 X(d) and Y(d)). This strain mismatch is a direct macroscopic measure of internal stresses.

For the field applied perpendicular to the cycling direction, the 90° domain switching that originates from the cycling direction shows this asymmetry (Fig. 2.12 X(c)), while the the other direction behaves just like the material had never seen any fatigue.

The crucial message from all the measurements of anisotropy is that fatigue only affects certain domains for certain orientations of grains. Fatigue induced defects thus must also be highly oriented in the material (see Chap. 3). Furthermore, the fact that 90° switching between both perpendicular directions is unaffected by the fatigue shows that only a fraction of the domain system participates in the switching process altogether. This fraction depends on the projection of the external electric field onto the domain system in a particular grain.

2.5 Leakage Current, Sample Coloring, and Relaxation

Leakage Current:

PZT is usually a highly insulating material at room temperature, unless significantly doped by hard dopants. Some changes in the leakage current may occur due to fatigue, because the charge state of several defects may change. So the leakage current was previously observed by Mihara et al. (1994a) to strongly depend on fatigue in the logarithmic regime. Their films of $Pb(Zr_{0.4}Ti_{0.6})O_3$ (100 nm thick, Pt-electrodes) showed three different regions in the current-voltage characteristics, two ohmic sections at low (< 10 kV/mm) and intermediate voltages and a highly non-ohmic current at voltages exceeding 100 kV/mm. The current densities at intermediate voltages significantly dropped with cycle number. The loss in current density with cycle number was accompanied by a reduced oxygen content of the samples measured via Auger electron spectroscopy. A significant reduction of E_c was also encountered, which differs from most other data.

In our data no significant increase in leakage current was observed. The current 2 to 3 hours after application of a d.c. field was still higher in a fatigued sample than in an unfatigued one. But this effect was uniquely due to the very slow relaxation currents, which still dominated the current across the sample after many hours. After about 5 hours the current in a fatigued sample ($3 \cdot 10^7$ cycles, 2 E_c) reached the leakage current density of 2 nA/cm^2 (discs of 1 mm thickness, 1.3 kV), identical to an unfatigued sample, but which already reached this value after 30 minutes (see relaxation beneath).

Sample Coloring:

Despite the fact that no macroscopic leakage current was observed in our samples, the local conductivity must have changed due to fatigue. As was observed optically, the samples darkened at a fairly early stage of fatigue as well as under sunlight. Thus charge carriers, which were initially bound to some trapping center were lifted into less deep traps, from which they can be thermally activated, become mobile and darken the sample. Coloring approaching the color of hard doped PZT ceramics was observed. This is equally obtained by UV light or by the cycling procedure. During later stages of fatigue the samples also darkened due to the increasing microstructural damages from microcracking.

Relaxation:

The data set that we have presently available on the effects of fatigue on relaxation of the domain system is very limited, but significant changes are obvious. One particular aspect has to be considered initially. Figure 2.13 shows the value of the coercive field in La-doped tetragonal PZT. It is extremely rate dependent. For very slow measurements of the hysteresis cycle,

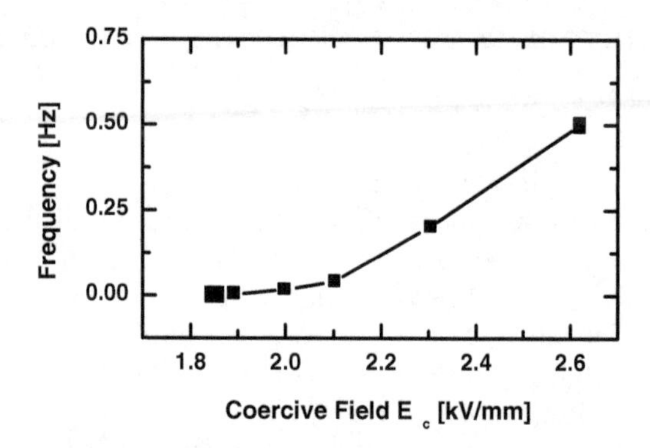

Fig. 2.13. Rate dependence of the apparent coercive field in fine grained tetragonal PZT (2%La)

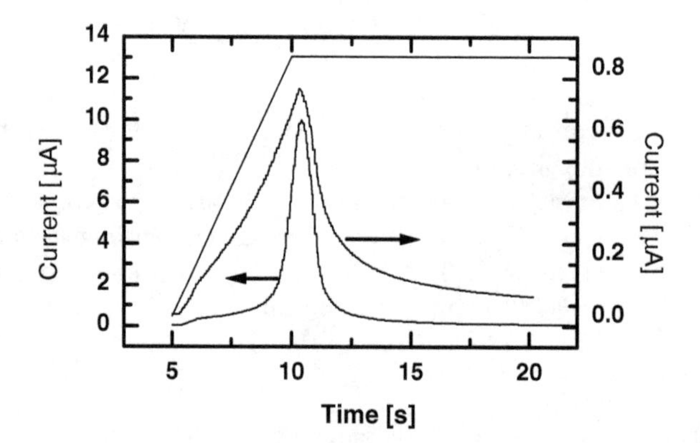

Fig. 2.14. Switching current relaxation of PZT before (left scale) and after (right scale) bipolar fatigue ($2 \cdot 10^7$ cycles). The externally applied voltage ramp is shown (max value $= 2\,\mathrm{kV/mm}$)

the measured values of the coercive field are low and increase with increasing measurement rate. Thus tetragonal compositions have a high tendency to exhibit long relaxation times for parts of their domain system. In rhombohedral ceramics no such effects have been observed.

Figure 2.14 shows the switching current measured in PIC 151 before (left scale) and after fatigue (right scale) for the voltage ramp indicated ($E_{max} = 2E_c$). It is clear that the maximum current drops by an order of magnitude and the maximum occurs at a later time. The entire fatigue story can probably be back-traced to rate effects. As will be shown in Chaps. 3 and 5, a dominant fatigue mechanism is the point defect agglomeration. The growth rates of the agglomerates can explain the loss in switchable polarization. On the other hand for very long times there is still some relaxation current. Thus large amounts of the domains are just delayed in their contribution to the switching charge. Domain wall motion across significant energetic barriers like extended defects is most probably thermally excited. In this case, larger barriers will be surpassed at a much lower probability and consequently at a much slower rate. This aspect of transferring static results on domain blocking to domain wall dynamics is not attempted in this work, but is the subject of further research.

As one aspect of the fatigue effect in PIC 151 is the change of the crystal structure from rhombohedral to tetragonal (see Sect. 3.3), the increased relaxation rates may also be partially due to the change in crystal structure. The above mentioned slow relaxation of the tetragonal compositions is then responsible for part of the effect.

2.6 Unipolar Fatigue

For the unipolar fatigue the setup of Sect. 2.1 was used. The effect of unipolar cycling is considerably different from bipolar cycling (Verdier et al., 2003). Figure 2.15 shows the polarization and strain hysteresis after 10^7 fatigue cycles. The second measuring cycle is displayed. The polarization hysteresis has slightly decreased in height. This means that the 180° switchability has decreased due to this cycling procedure. The strain is also asymmetric like in the case of bipolar cycling. Nevertheless, there are two competing effects here. The asymmetry is high, indicating the formation of a bias polarization like in the samples after bipolar cycling, but the total strain of the larger of both wings has increased. This is definitely different from the bipolar case. For this case of change in material behavior the term degradation may be slightly misleading, because there seems to develop no definite loss of switchability of the domains. There is just a definite imprint of an offset polarization. Again, this differs from an offset field, which is observed after extended d.c. voltage loading of perovskite type ferroelectrics. Actually, both effects are induced after unipolar cycling, the bias field and the offset-polarization. The first hysteresis is slightly shifted along the field axis. This effect is directly removed by the first bipolar measuring cycle. Thus, d.c. loading at constant field and unipolar a.c. cycling yield significantly different material behaviors, because the offset-polarization is a fairly stable property of the material after the unipolar fatigue. It is possible that the same is true in samples that were long subjected to pure d.c. loading, but strain was never tested. Thus a combined effect of an offset-polarization and an offset field may also be present there. Their relative contributions may considerably differ in importance though.

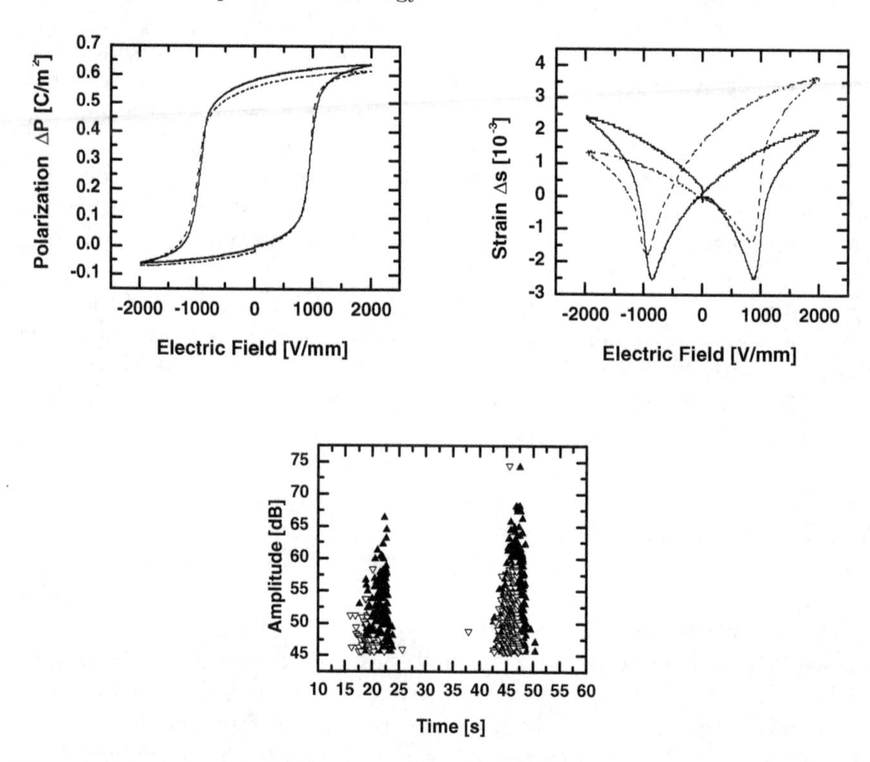

Fig. 2.15. Polarization (left) and strain (right) hysteresis, and acoustic emissions (see discussion in Sect. 4.2.3) before and after (dotted line) unipolar electric loading at 2 kV/mm, 10^7 cycles. The open symbols are acoustic emissions after unipolar fatigue plotted versus time for a bipolar triangular voltage ramp

2.7 Mixed Loading Fatigue

The setup for this experiment was described above in Sect. 2.1. Samples were cycled up to $6 \cdot 10^6$ cycles. Figure 2.16 (uncracked sample, see beneath) shows the strain and electric hysteresis loops after $2 \cdot 10^6$ cycles. The polarization hysteresis has only slightly dropped, while the strain hysteresis has developed a marked asymmetry. The asymmetry after mixed loading is similar to unipolar fatigue shown in Fig. 2.15. In both cases the fatigue essentially induces a polarization offset without a marked decrease of the absolute amplitude of the hysteresis. This is a fairly remarkable result. While the bipolar fatigue induces a reduction of the polarization hysteresis of more than 50% the mixed loading hardly reduces the switchability of domains in PZT. The cycling conditions were intentionally chosen such that significant back-switching occurs in PZT because the compressive coercive stress (40 MPa) was exceeded by more than 30%. Thus, it is clear from this experiment, that the essential driving force for the harsh reduction in polarization fatigue is driven by 180° switching. This is fairly remarkable, because the 180° switching is usually

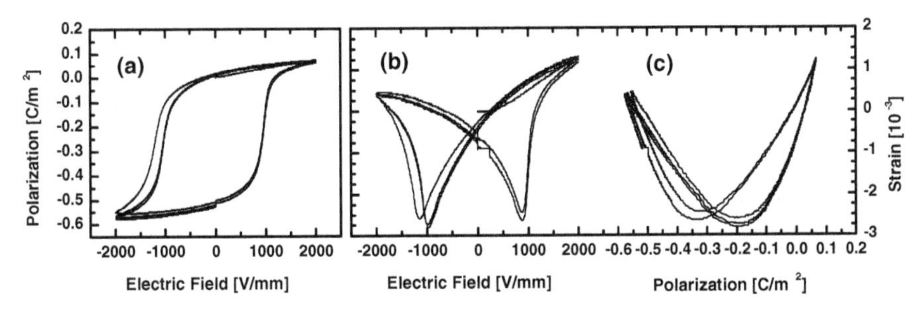

Fig. 2.16. Polarization (**a**), strain (**b**), and electrostrictive (**c**) hysteresis after $2 \cdot 10^6$ cycles mixed unipolar electric (2 kV/mm) and compressive mechanical (55 MPa) loading. Several consecutive measurement cycles are shown. No mechanical load is applied during measurement

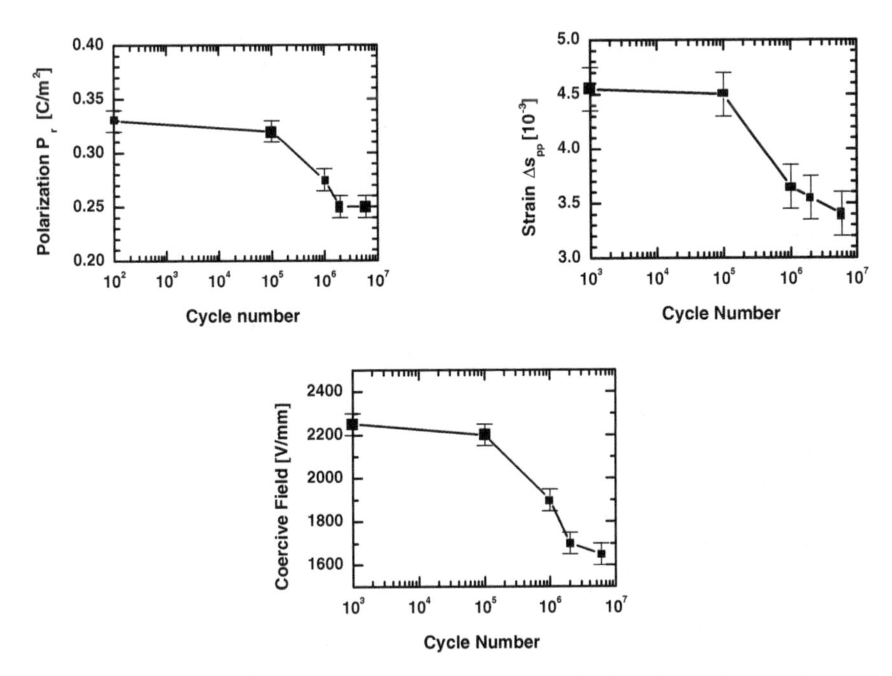

Fig. 2.17. Change of material parameters after mixed loading fatigue, left: Remnant polarization, right: Peak-to-peak strain in the more degraded hysteresis wing, and bottom: Coercive field

considered the easier switching type, inducing e.g. much less acoustic emissions (see Chap. 4 beneath) and is thus expected to much less interact with defects. This can thus not be true for at least some defects.

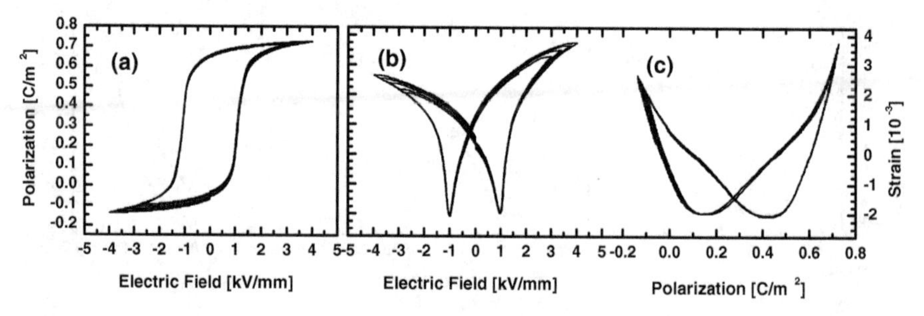

Fig. 2.18. Polarization (**a**), strain (**b**), and electrostrictive (**c**) hysteresis after $2 \cdot 10^6$ cycles mixed unipolar electric (2 kV/mm) and compressive mechanical (55 MPa) loading. The field applied to the sample during measurement is increased in consecutive cycles. No mechanical load is applied during measurement

If subjected to high fields, the polarization hysteresis hardly increased in amplitude, while the strain grew significantly on both of the asymmetric wings. Different from the results of high field treatment in samples after bipolar fatigue (Figs. 2.7 and 2.8), the asymmetry remained essentially unchanged at these high fields. This is an indication that unipolar mixed loading fatigue induces a much more stable offset-polarization than bipolar fatigue. On the other hand the absolute strain values hardly dropped due to unipolar and mixed loading fatigue. Another aspect is crucial when observing the hysteresis data in all three cases. The bipolar electric cycling induces a significant slanting of the polarization hysteresis resulting in a broadening of the strain hysteresis minimum around E_c. In the unipolar and mixed loading fatigued samples this strain minimum is still sharp. Thus, the relaxation behavior, or in other words a distribution of coercive fields is only induced after bipolar fatigue.

Another set of relevant observations on the mixed loading experiments was the formation of cracks. As mentioned in the experimental section, the strain mismatch between the rods providing the external loads and the samples was hoped to be matched by using metal pieces of approximately the same Poisson's ratio (0.36) and elastic constant as that of PZT. It was hoped that the change of transverse strain due to the ferroelastic switching, which actually translates to a Poisson's ratio of 0.5 (Cao and Evans, 1993) would be insufficient to fracture the samples, because it would primarily also lead to compressive stresses. This held true up to 10^5 cycles. Then the samples started showing small surface cracks at 10^6 cycles growing into a crack system on the entire surface of the sample at $6 \cdot 10^6$ cycles. To estimate, whether this fatigue crack growth stems from fatigue of the material itself or from the boundary conditions, three samples were placed as a sandwich into the setup (evidently the triple voltage was applied). Fracture was actually due to the strain mismatch between the samples and the metal pieces. The center samples did not fracture in a single experiment and the "bread in the sandwich only showed cracks in the crust". The observed crack depths were only in the upper half of the samples adjacent to the metal clamps. Only single samples not part of a sandwich showed through thickness cracks. The strain and polarization hysteresis shown above were taken from center samples in order to display the pure material effect.

When considering measured transverse strains in soft PZT's, they are complex functions of the strain state within the stress strain-hysteresis and change the effective values of Poisson's ratio from 0.4 to 0.5 in the material used by Cao and Evans (1993). The cracking in our samples is thus a fatigue effect induced by changing the Poisson's ratio during the mixed cycling procedure and clamped by the neighboring metal blocks. The cracking phenomenon seems somewhat more complex, though, and will have to be addressed in further work.

3 Agglomeration and Microstructural Effects

3.1 Agglomerates

One of the major results of this work is presented in this section. It is a very simple finding with many implications. Upon the etching of a fatigued sample (2 E_c, 10^8 cycles) with an HF/HCl mixture (1.6 ml HF 49%, 2.6 ml HCl 37%, filled to 200 ml) the domain patterns (2-6 seconds), the grain boundaries (2-20 seconds) and some additional etch pits appear (60 seconds). The particular shape of the etch pits is shown in Figs. 3.1 and 3.2.

The most commonly encountered type is depicted in Fig. 3.2(c). Etch patterns mostly appear within single grains or sometimes crossing single grain boundaries. They are mostly linear or lengthy and will thus be called etch grooves (previously called "corrosion paths" (Nuffer et al., 2000)). As the acid attack is very harsh, the original structures of the defects facilitating the acid attack are etched away. Nevertheless, the linear character of the grooves is apparent. The total amount of such etch grooves is highest in the center of the sample half way between both electrodes as can be seen from Fig. 3.3. Regions near the electrodes only seldom contain etch grooves.

The etch grooves are most likely planes in the crystal that are highly charged. Linear arrangements are also possible according to the models discussed in Sect. 5.8. Their intersections with the polishing plane would in most cases only yield single spots like etch pits generated from dislocations when these cross a polished surface (Reed-Hill and Abbaschian, 1994). Extended linear patterns have to originate from planes intersecting with the surface. As will also be developed in Sect. 5.8, an assignment of these etch grooves to planar point defect agglomerates is given based upon electrochemical agglomeration models like the one proposed by Brennan (1993) and Dawber and Scott (2000a). According to Brennan (1993) the interaction of the domain walls with the localized defects necessitates planar arrangements, if defects do cluster in perovskites (see also p. 36 in the introduction).

3.2 Grain Boundaries

Another effect was observed near the electrodes, where the etch grooves are rare. Figure 3.4 displays a TEM image of a fatigued sample near the elec-

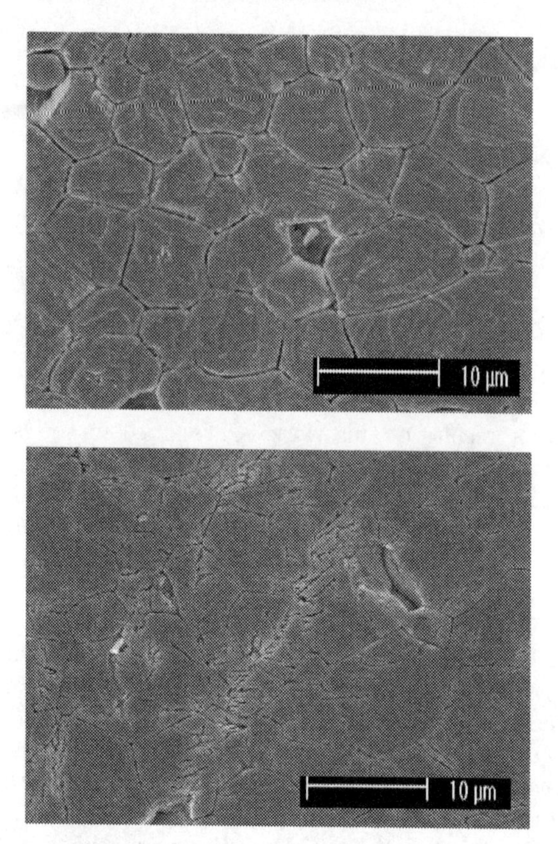

Fig. 3.1. Etch grooves in fatigued PZT after 60 seconds etching with HF/HCl, polished parallel to the electrodes, view along z-axis (see Fig. 1.5). (**a**) Unfatigued sample, top and (**b**) fatigued sample with dendritic tree of etch grooves, bottom

trode surfaces at a depth of about 50 µm. The material in triple points has disappeared, while in the depth of a fatigued sample and in unfatigued samples the triple points also contain some material after ion thinning. To our understanding the triple points closer to the electrodes are weaker and the material is washed out during the ion thinning of the sample preparation for the TEM, while regular triple points of an unfatigued material or in the center of the sample are sturdier. The fatiguing procedure has thus produced some changes in the structure of the triple points, but only in the electrode proximity. The TEM analysis also revealed that the grain boundary contains some amorphous material. This was not observed prior to fatigue, when the grain boundaries were intact and no obvious amorphous layer was detected. So far, we have not been able to uniquely assign a microscopic mechanism for this amorphization of the grain boundary triple points.

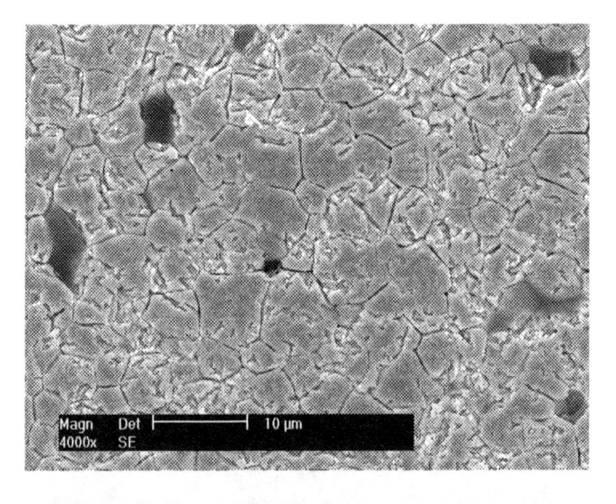

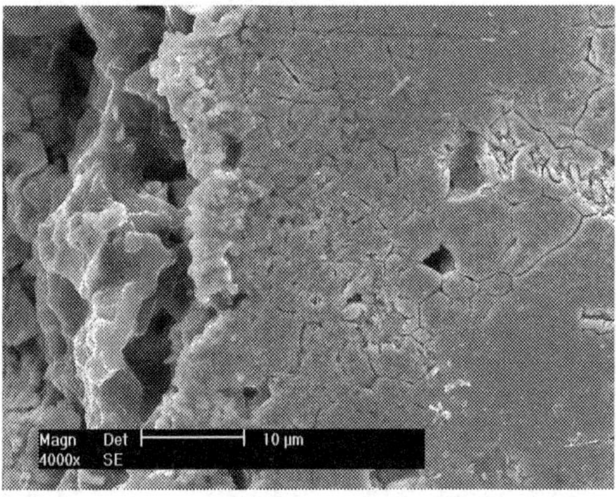

Fig. 3.2. Etch grooves in fatigued PZT after 60 seconds etching with HF/HCl viewed perpendicular to z-axis (see Fig. 1.5). (**c**) Common etch groove pattern seemingly confined to single grains, top and (**d**) etch pattern underneath the electrodes, bottom

A first point of interest with respect to fatigue and the profiles found in this study is the behavior of the silver paste electrodes. A detailed analysis by Nagata et al. (1997) employing secondary ion mass spectroscopy (SIMS) suggests, that at the firing temperature of the electrodes in our study (120 min, 720°C) the major diffusion of silver occurs along the grain boundaries. The diffusion of silver into the bulk grains of PZT on the other hand is very limited. Nagata et al. (1997) determined the following volume D_{v} and grain boundary D_{gb} diffusion constants:

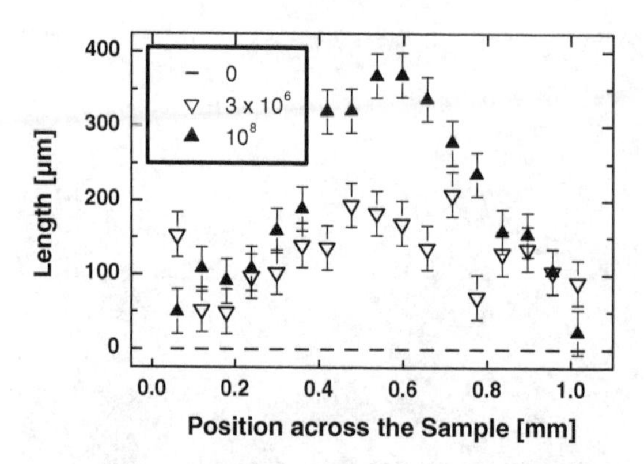

Fig. 3.3. Distribution of etch grooves across the sample profile between both electrodes (left and right). Displayed is the total length per area ($60 \times 40\,\mu m^2$)

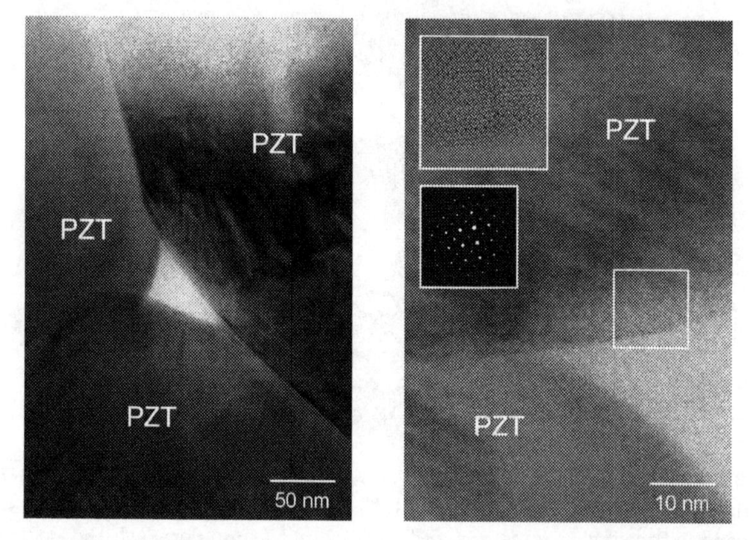

Fig. 3.4. TEM image of a triple point after 10^8 fatigue cycles at $2E_c$ near the electrodes (about 40–50 µm, depth). On the right side the transition from the crystalline grain to an amorphous layer is shown. The center of the triple point is empty

$$D_{\mathrm{gb}} = 43.5 \exp\left(\frac{(-236 \pm 46)[\mathrm{kJ/mol}]}{RT}\right) \quad \left[\frac{\mathrm{cm}^2}{\mathrm{s}}\right] \tag{3.1}$$

$$D_{\mathrm{v}} = 0.12 \exp\left(\frac{(-258 \pm 25)[\mathrm{kJ/mol}]}{RT}\right) \quad \left[\frac{\mathrm{cm}^2}{\mathrm{s}}\right] . \tag{3.2}$$

In the case of thermal diffusion into the material,

$$D_i = \frac{\langle(\Delta x)^2\rangle}{2t} , \tag{3.3}$$

the average length scale of the Ag-diffusion into the ceramic is $\Delta x_{gb} = 11\,\mu m$ and $\Delta x_v = 0.02\,\mu m$. Particularly the diffusion along the grain boundaries is thus significant and can reach far into the sample. Values of the diffusion constants given in another publication are even 3 to 5 orders of magnitude higher (Slinkina et al. (1993), see p. 68).

Furthermore, Nagata et al. (1997) showed that regions, where silver had diffused into, had much lower grain boundary strength than the bulk PLZT during fracture. Right at the border between areas that were penetrated by some silver and those that were free of silver, the fracture surface changed from intergranular to transgranular. Also the chemical etching resistance (48 h, nitric acid, unspecified concentration) of the grain boundaries (Ag-firing, 6h, 700°C) was reduced due to the silver and yielded well visible etch patterns of the grain boundaries in the silver enriched areas.

The second pertinent point is the enhanced amount of lead oxide in the grain boundaries, particularly in the triple points as was found by Hammer (1996), at least in the case of La-doped PZT. Lead oxide itself is mechanically much less stable than the perovskite material. In the unfatigued case though, the material is strong enough to survive the ion thinning, while some extra effect has to occur during fatigue, that weakens the grain boundary.

Results from the Karlsruhe group (Hammer et al., 1998) also showed that silver (Ag^+) can be incorporated into the PZT structure as an acceptor ion. This correlates well with the findings by Nagata et al. (1997) that the lead content was reduced in those regions, where silver had diffused into. The very volatile PbO is substituted in the grain boundary regions by some silver compound or metallic silver and the PbO is driven out of the sample. An important aspect of the latter mechanism is the oxidation of silver, for which the necessary potentials have to be provided either during the firing of the electrodes or during the bipolar fatigue.

The data for the near electrode regions by Nagata et al. (1997) also suggest, that not only grain boundary material is substituted. For the first one or two grain layers severe loss of lead is also encountered within the grains. For the defect reaction

$$2\,\mathrm{Ag} + 2\,\mathrm{PbTiO_3} + \mathrm{O_2} \longrightarrow 2\,\mathrm{AgTiO_3} + 2\,\mathrm{PbO} \qquad (3.4)$$

the oxidation of silver is necessary to incorporate it as an acceptor ion into the PZT structure ($PbTiO_3$ chosen to show the effect). Furthermore there has to be a sufficiently open structure of the grain boundaries to permit the diffusion of atomic oxygen, silver and lead oxide.

In the depth of the sample no free oxygen is present and oxygen vacancies have to be formed during the similar reaction

$$\mathrm{Ag} + \mathrm{PbTiO_3} \longrightarrow \mathrm{AgTiO_2}(V_O^{\bullet\bullet}) + \mathrm{PbO} + 2\,e' \,. \qquad (3.5)$$

Considering that lead is oxidized much easier than silver, both reactions are fairly unlikely unless high electric fields are present to accommodate for the charge imbalance. For the latter scenario still a sufficient diffusion of lead oxide has to be present to allow for a net decrease of the lead content as found by Nagata et al. (1997).

In a second work by Slinkina et al. (1993) the contents of silver diffused into the depth of the sample were quantified using radioactive tracers. The maximum average solubility of silver in polycrystalline PZT was determined to be $1.4 \cdot 10^{19}$ $1/\text{cm}^3$ at 720°C after more than 60 h. After 1 h of heat treatment, the maximum value near the electrode is $1.0 \cdot 10^{19}$ $1/\text{cm}^3$ and drops down with the expected diffusion profile to $0.15 \cdot 10^{19}$ $1/\text{cm}^3$ at 500 µm depth. These values are reduced for Ag/Pd alloys by about a factor of 3-4. The average concentrations have to be corrected for the effects shown by Nagata et al. (1997). Considering an average thickness of the grain boundary phase of 10 nm and 5 µm grain size, the number of unit cells occupied by Ag-atoms rises to 2% of all unit cells in the grain boundary at saturation. This concentration is far from the percolation limit (0.312 simple cubic, 0.245 BCC, body centered cubic, and 0.198 in FCC, face centered cubic, lattices (Stauffer, 1985)). In the real device the concentrations are of this order of magnitude only at the electrode and drop down considerably towards the sample interior.

Qualitatively the diffusion data by Nagata et al. (1997) and Slinkina et al. (1993) agree reasonably well with the profiles found in our samples, but the absolute silver content is not clear here. The lack of defect agglomerates in the regions near the electrodes (Fig. 3.3) is most likely associated with the incorporation of silver into the material with the grain boundaries being the dominant diffusion path. As the SIMS as well as nuclear methods are only sensitive to the ion itself, the charge state of the silver in the grain boundary or the grain surface is not clear. Most likely some of it will be oxidized due to the high local electric fields during fatigue and some will still be metallic. The differences in Ag-diffusion under positive and negative electric field suggest that some silver is charged and was thus oxidized during the electrode firing and subsequent application of electric field.

Three possible results for the fatigue behavior can be derived. The first argument is along the lines of earlier fatigue pictures by Scott et al. (1991). Conducting dendritic percolating paths exist along the grain boundaries due to the diffused silver. This entails regions, where the average electric field is highly reduced, while local fields at certain dendrite end-points become extremely high. On average, some area close to the electrodes will not experience any fields at all, while regions towards the end of the diffusion range of the silver will locally experience fields much higher than the average value. In such a case the electrically induced strain mismatch will lead to very high local stresses and possibly to the immediate formation of microcracks in a certain part of the sample. This scenario is unlikely though, because a change of crystal structure (see Sect. 3.3 beneath) is also observed directly at the electrodes, implying that these regions do experience electric fields.

The second result may be that the oxygen vacancies as the charged mobile species drift into the grain boundary region under the high local fields. With a high number of vacancies present, the grain boundary region weakens. In this scenario, the lead oxide in the grain boundary region decomposes to lead and the perovskite is deprived of mobile charge carriers, or actually atomic oxygen diffuses into the material. This also corresponds well with the lack of finding the types of etch grooves from Sect. 3.1 in the regions close to the electrodes. The etch grooves are the result of a local rearrangement of the oxygen vacancies. In the depth of the sample no exchange with the exterior is possible in reasonable amounts of time, while near the electrodes the diffusion of the vacancies is highly driven into the partly conducting grain boundaries and dominating over local rearrangement in the grains. If a dominating part of the silver ions is already oxidized, similar arguments still apply for fatiguing over many cycles. As silver is an acceptor ion in PZT (Xu, 1991; Hammer, 1996) it strongly enhances the conductivity of the PZT which is a well known effect for hard doped materials (Xu, 1991). Thus the local drops in potential in the regions near the electrodes will still be lower than in the interior of the sample. This is actually the opposite effect as in thin films, where space charge layers form underneath the electrodes that exhibit much lower dielectric constant than the ferroelectric material and thus higher potential drops occur across this layer and the sample interior is partially screened from the external field (see Sect. 1.10).

The third scenario permits a certain agglomeration of metallic silver, but no percolating paths. This scenario will induce enhanced and reduced local fields for certain locations. It has the advantage that a percolation across the sample is effectively not possible, while it can result from a scenario of the first type.

The second scenario is the one favored from my point of view. The weakening of the triple points is easily explained. Furthermore, no indications of an inhomogeneous distribution of microcracks was found, even though an exhaustive statistical account of the number of microcracks was not feasible using the TEM technique in reasonable time.

The present conclusion is the following:

In the regions close to the electrodes, the grain boundaries constitute paths of partial conductivity. Oxygen vacancies under applied electric fields are drawn to the grain boundaries and weaken this part of the microstructure. In the depth of the sample no conducting paths are present. The charged defects agglomerate to planar structures within the grains, where they are originally located after sintering. This agglomeration is responsible for the strongly reduced mobility of the domain system in this part of the sample.

This conclusion will be further supported in the next section.

3.3 Agglomeration and Crystal Structure

The interaction of the defect agglomerates with the domains is not arbitrary in direction, because the agglomerates themselves occupy well defined {100} planes (Suzuki et al., 2000), which is a necessary condition by the perovskite type structures. In the case of the tetragonal crystal class two out of possible six domain orientations perfectly accommodate the electric fields generated by a charged plane.

Arlt (1990b) showed a complete scheme of 90° domain walls to fill up space. Within this three dimensional structure certain 180° domain walls may be charged as delineated in Fig. 5.5. Unfortunately, these charged walls do not occupy {100} planes. Thus, beyond the fact that the planar agglomerates exert forces on rhombohedral domains, they also do not match the only charged domain walls possible in a three dimensional filling of space by domains in the tetragonal structure. A reduction of the electric potential energy is not directly possible. Also in the tetragonal case the formation of additional domains is necessary, as will be described in detail in Sect. 5.8.2.

In the rhombohedral class the potential energy of none of the polarization directions is zero with respect to the {100} dislocation loops. Any polarization will form an angle of either 35.3° or 54.7° (Fig. 5.3 (f)) with the agglomerate plane. Consequently, there will be a different energy maximum associated with 109.4° or 70.6° domain walls when they try to surpass the dislocation loop. The acoustic emission energies associated with the discontinuous motion of 90° (109.4° and 70.6°) domain walls should thus differ between the crystal classes for the identical defect, but such an effect will be small and hard to prove.

Another implication of the misorientation of the rhombohedral domains with respect to the planar dislocation loops is the energy stored in the domains themselves. For all domains not oriented at 90° to the defect plane a shear stress is exerted onto the domain. If the defect plane is sufficiently large and the external constraints on the rhombohedral grain are moderate, the elastically stored energy due to this shear stress may exceed the energy needed to switch the domain from rhombohedral to tetragonal. This seems actually to be the case in PIC 151. In Fig. 3.5 a larger number of supposedly tetragonal needle domains was observed in the microstructure after fatigue. A larger set of photographs was taken in the SEM after etching. In certain samples the entirely rhombohedral structure switches to entirely tetragonal after the fatigue of $2 \cdot 10^7$ cycles at 2 E_c. The phase switching occurs first in the center of the sample. Then the number of tetragonal grains increases and grows towards the two electrodes. After $1 \cdot 10^7$ at 2 E_c around one half of the microstructure has changed from rhombohedral to tetragonal. At the end of the fatiguing also grains directly underneath the electrodes switch to tetragonal. Thus the entire sample does experience fatigue of individual grains and thus electric fields, and the phase switching occurs all the way to the electrodes. The agglomeration to planar defects therefore also occurs near the

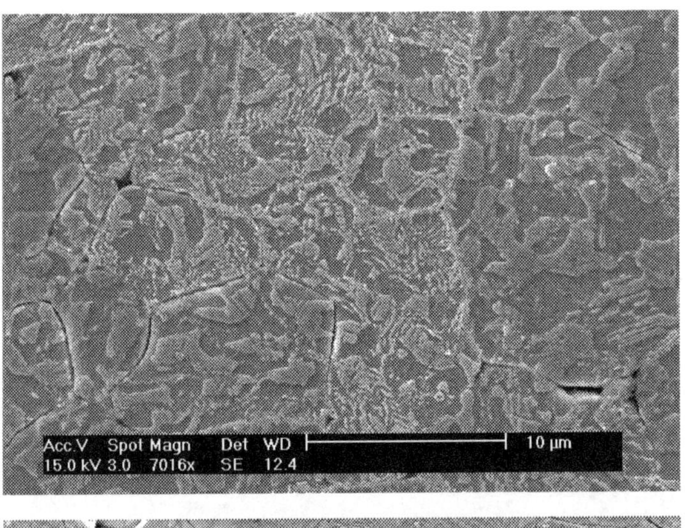

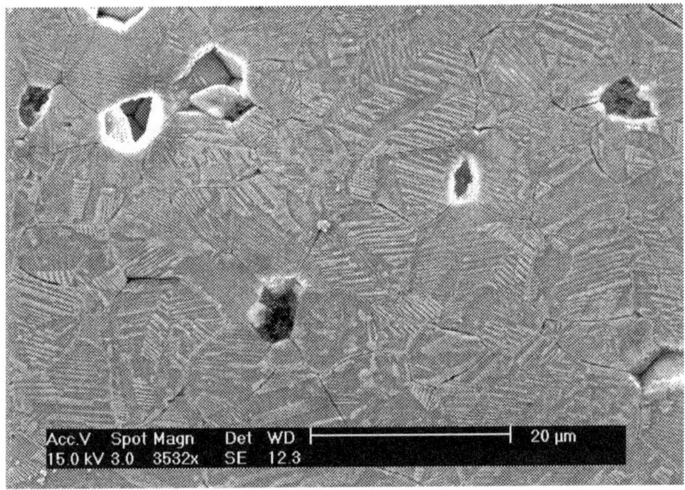

Fig. 3.5. Unfatigued and fatigued ($2 \cdot 10^7$ cycles) microstructure of PIC 151. In this particular batch of samples the crystal structure changes entirely from rhombohedral (top) to tetragonal (bottom)

electrodes, but at a later stage of fatigue and to a lower degree. This suggests that the local fields in the grains are lower in the proximity of the electrodes. To describe fatigue as a phase transition itself was recently proposed by Scott (2001). There are 2 other interpretation of the altered domain pattern. The first one is the massive formation of needle domains due to the agglomerates in the bulk (see Sect. 5.8.4), a second one concerns the increased ionization of local defects and a subsequent breakup of larger domains into smaller entities, which entails significant changes of domain dynamics (a discussion of this approach is given elsewhere: Lupascu et al. (2004)).

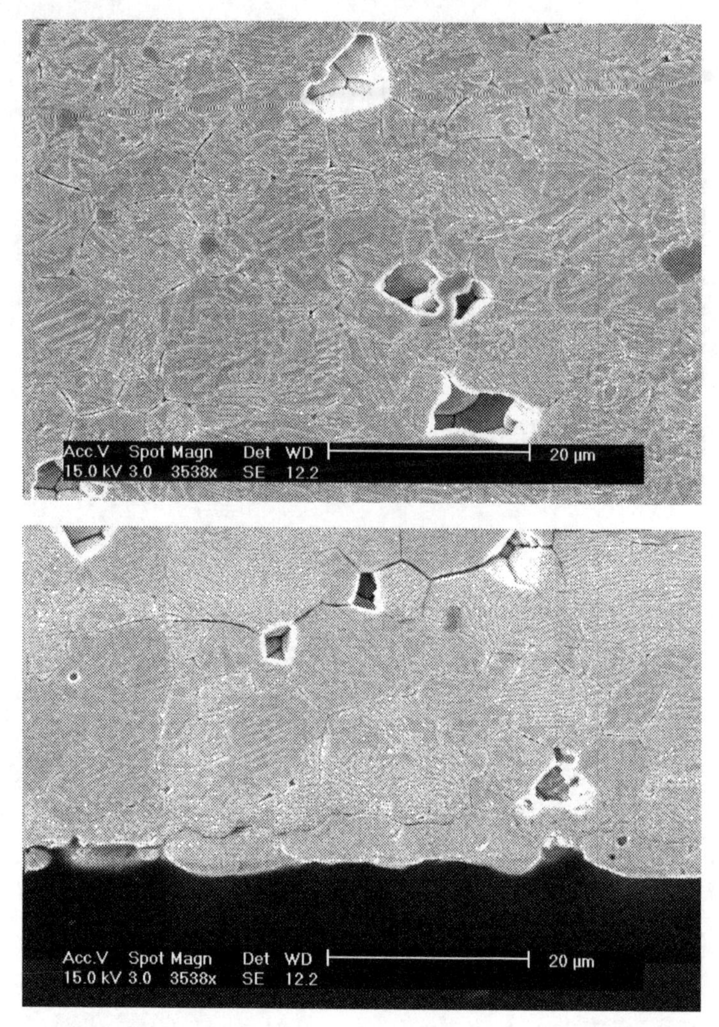

Fig. 3.6. Crystal structures at an intermediate cycling stage at different positions in the sample. Center of the sample (top) and near the electrode (bottom)

3.4 Domain Structure

In Fig. 3.7 two images of the domain system are taken before and after fatigue (Lupascu and Rabe, 2002). It is evident that the domain structure changes entirely. The initially watermark like domain pattern of predominantly 180° domain walls changes drastically into a high density system of 90° domain walls occurring throughout the sample.

Fig. 3.7. Domain structures (**a**) before and (**b**) after 10^8 bipolar cycles at 2 kV/mm

3.5 Unit Cell Volume

In Fig. 3.8 the unit cell volume is shown across the sample. Irrespective of position the unit cell volume decreases with cycling. Pan et al. (1992c) report in their work, that the degree of tetragonality increases with cycling, the c-axis length of 0.204 nm in the thermally depoled state grows to 0.2053 for the fatigued sample, with equal a-axis length in both cases, which is contradictory to our data. Their interpretation on the other hand corresponds to our one. They attribute the shift in their peaks more to a fitting artifact, because the rhombohedral peak has to be included. They assign this effect also to a higher number of tetragonal grains in the ceramic after fatigue (see Sect. 3.3), which can also explain their fits.

A way to explain the reduced unit cell volume is that an increased number of acceptor doping leads to a reduced c/a-ratio of the unit cell (Kamiya et al. (1993), the authors could actually correlate this change in lattice constant to variational cluster calculations, from which the changes in energy of the substitutional defects could be determined). During fatigue the darkening of our samples was observed, an effect also occurring in soft doped materials under UV-light illumination (see Sect. 5.3.3). A similar effect occurs in acceptor doped materials, in each case due to the enhanced number of free carriers. Thus some ions in our material will change their charge state, possibly Pb^{3+}, which has been well described by Warren et al. (1994b) (again Sect. 5.3.3). Such large numbers of highly ionized defects may induce the reduced average

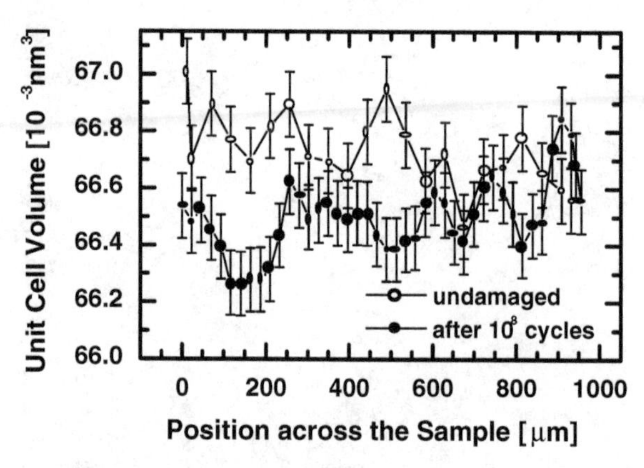

Fig. 3.8. Tetragonal lattice cell parameters across the sample profile before (open circles) and after (closed circles) fatigue (10^8 cycles, $2\,E_c$)

unit cell volume. Other models involving clamped domain configurations are more complicated and will not be attempted here.

3.6 The Oxygen Balance

It has been a long standing belief, that oxygen is also liberated from bulk samples in larger amounts. This may in fact be true for a very thin layer at the surface. For the deeper regions in the bulk the diffusion constants are much too low to drive considerable amounts of oxygen to the sample surface. This fact is verified in mass spectrometer measurements shown in Fig. 3.9 under extremely low gas fluxes (Nuffer et al., 2001).

Large amounts of oxygen are observed to leave the sample container under bipolar cycling. The measurements have to be performed in an insulating liquid environment. At reduced pressures of nitrogen or argon instead of a liquid insulator the electron emission from the sample surface immediately triggers the formation of a plasma. As can be seen from the measurement in Fig. 3.9, the liberation of oxygen is always accompanied by liberation of nitrogen. The amounts of oxygen and nitrogen dissolved in the silicon oil were calculated according to known solubility data and well match the amounts freed after long term cycling. In Fluorinert® the solubility of both gases is even higher than in silicon oil. The sonic excitation, which occurs due to the switching of the sample, drives these dissolved gases out of the liquid. This was reproduced by simply placing the silicon oil into an ultrasonic bath. The amount of oxygen that can still escape the sample when the amounts stemming from the oil are subtracted and the errors in the experiment considered is less than 10^{-4} of all oxygen atoms in the sample. Thus less than 10^{-4} of all oxygen can escape the sample at most. Considering the diffusion data for

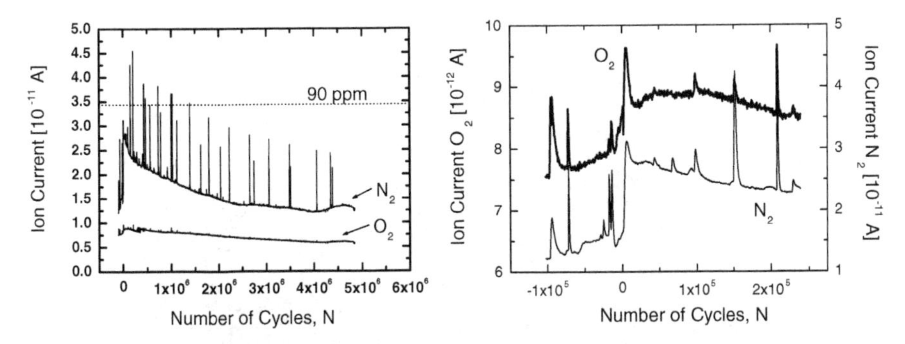

Fig. 3.9. Comparison of gas flow from PZT sample during cycling immersed in silicon oil. Negligible oxygen liberation from PZT, but ultrasonically pumped gases from the insulating silicon oil. Detail shown to illustrate the coincidence of N_2 and O_2 gas liberation on the right

oxygen, these numbers have to be reduced even more. Thus the data by Pan et al. (1996) have to be understood as the liberation of gases from the insulating liquid surrounding the sample during cycling and not from the crystal lattice.

Irrespective of the discussion above a local reordering of the oxygen vacancies within the microstructure is reasonable at cycle number commonly encountered to be responsible for fatigue (10^8 cycles). The mean distance travelled by oxygen vacancies will be in the 100 nm range at room temperature.

3.7 Microcracking

Microcracking in general is a phenomenon not fully understood in ceramics (Kingery et al., 1976; Vedula et al., 2001; Zimmermann et al., 2001). Particularly the stress values at which microcracking occurs are generally too low for rupturing the chemical bonds in perfect single crystals (Lawn, 1993). Microcracking thus mostly occurs at grain boundaries, which are generally weaker than the neighboring grains. But still some stress concentration mechanism has to be assumed to account for the stresses necessary to break chemical bonds and this is still subject of intense discussion.

In ferroelectric PZT a large portion of the local stresses may be due to the interaction of a domain wall with the grain boundary. In a theoretical work Zhang and Jiang (1995) determined the character of the singularity of such an intersection point of a wall with the grain boundary. Their analysis does not yield a very high concentration of stresses at the grain boundary for an uncracked sample. The singularity is of order $1/r^{0.1}$, which is comparatively weak with respect to the $1/\sqrt{r}$ dependence encountered for macroscopic

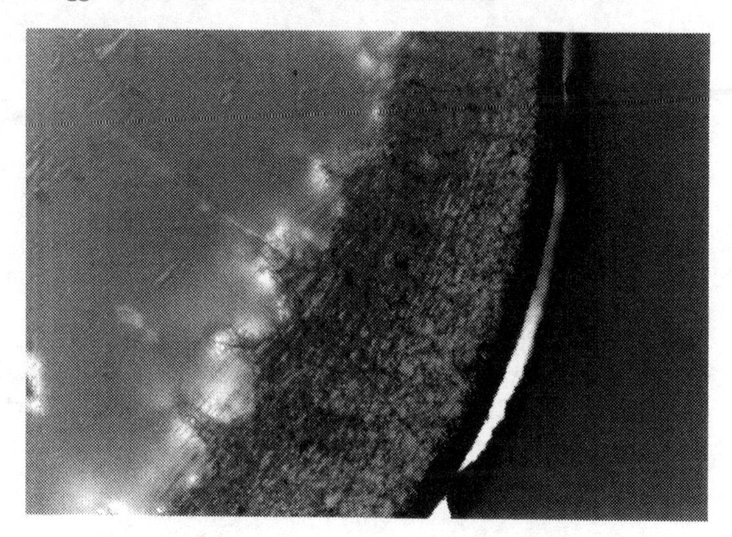

Fig. 3.10. Microcracking at electrode edges in transparent electrostrictive PLZT 9.5/65/35

cracks. If other microcracks already exist, the local stress singularity becomes more pronounced than for long cracks, a $1/r^{0.6}$ dependence is found. Altogether, there is no indication that in ferroelectrics the necessary stresses are easily determined from the material properties and similar problems are encountered as in other ceramics. Nevertheless, some stress enhancement has to take place.

3.7.1 Edge Effects

The easiest way to arrive at large local stresses is a singularity like an electrode edge, which will induce high local stresses due to a macroscopic constraint (Lucato et al., 2001a). Figure 3.10 shows an image of an electrostrictive transparent material (9.5/65/35 PLZT) after $3 \cdot 10^6$ switching cycles at 2 kV/mm near the electrode edge. Large amounts of microcracks are found. It is crucial in this experiment, that the microcracking is strictly mechanically induced. No ferroelectric switching occurs in 9.5/65/35 PLZT, but high strains are achieved at 2 kV/mm.

In ferroelectric materials we have performed several studies showing that mere poling induces a large amount of starter cracks at the electrode edge, even without any further cycling (Lucato et al., 2001b). These initial cracks then may serve as new concentrators for fields and stresses in their environment and facilitate further fatigue. A more detailed treatise of edge effects can be found in Lucato et al. (2001a).

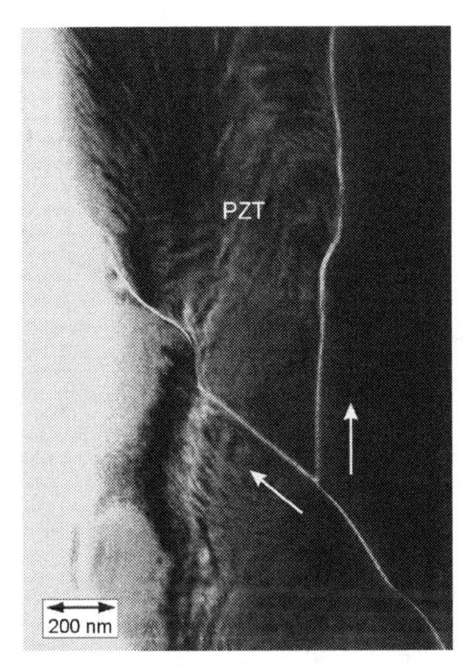

Fig. 3.11. A microcrack in the center of a bulk PZT sample of 1 mm thickness

3.7.2 Bulk Microcracking

A substantial study of microcracking phenomena was given by Kroupa et al. (1988, 1989). Their study showed that for sufficiently large grains in polycrystalline PZT the internal stresses due to poling are sufficient to generate microcracks. The microcracking is anisotropic with respect to the electric field direction yielding a higher number of cracks perpendicular to the field direction than parallel to it. The observed effects are explained in the framework of grains favorably oriented with respect to the electric field surrounded by unfavorably oriented grains. The strain mismatch induces the stresses necessary for the formation of the microcracks.

In our own samples of PIC 151 significant amounts of microcracking arise after fatigue. These occur homogeneous throughout the entire volume of the sample. An example found in TEM is shown in Fig. 3.11. Unfortunately, these studies are not possible using the SEM, because the crack opening displacement is not sufficient to be observed at the surface of a sample and be differentiated from ordinary grain boundaries. A statistical study using TEM is extremely time consuming due to the difficult sample preparation. Thus a statistical investigation correlating e.g. the change in crystal structure at intermediate stages of fatigue with the distribution of microcracks was not possible. Nevertheless, as the distribution of microcracks is homogeneous throughout the sample interior after 10^8 cycles, but the number of etch

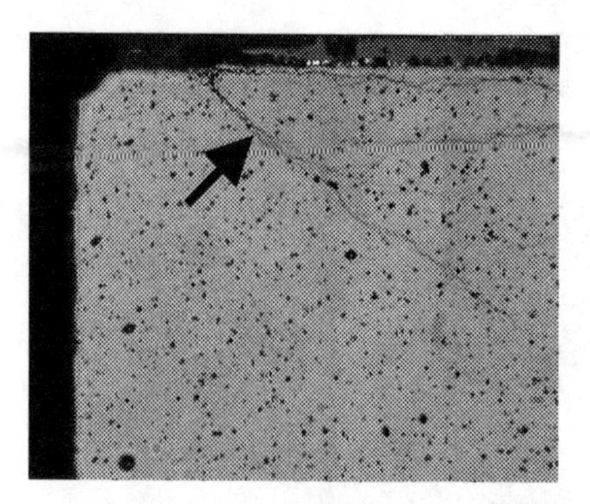

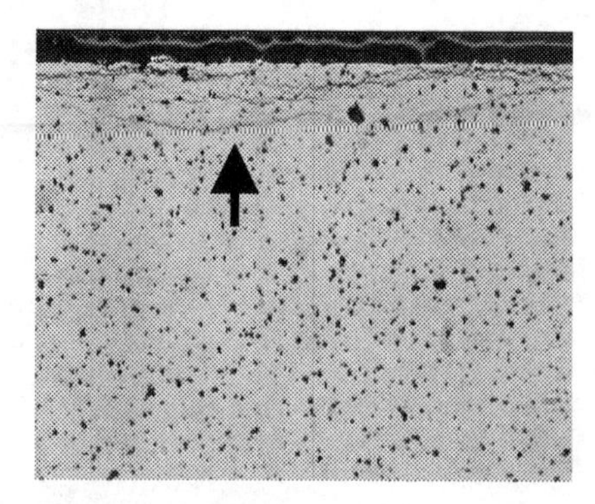

Fig. 3.12. Delamination cracks in bulk PZT (PIC 151, 1 mm thickness) after 10^8 bipolar cycles at $2\,E_c$. (left) Conical cracks due to the strain mismatch with the non electroded sample rim and (right) delamination cracks parallel underneath the electrode faces

grooves is not, the association of microcracks with the etch grooves will not capture the entire effect . Figure 3.3 well displays the much higher density of such grooves in the center of the sample particularly in the late stage of fatigue.

3.8 Macrocracking, Delamination Fracture

A well known form of cracking in ferroelectrics are cracks initiated at the electrode edges of cofired multilayer stack actuators, which several authors have reported (Winzer et al., 1989; Freiman and Pohanka, 1989; Furuta and Uchino, 1993, 1994; Schneider et al., 1994). It is due to the strain mismatch between regions experiencing high electric fields between the electrodes and regions remotely ahead of the electrode edge that are field free. This mismatch generates tensile stresses perpendicular to the electrodes in the field free region, which causes fracture ahead of the electrode edge. At the electrode edge itself electric fields become singular and additionally generate volumes of extreme strain mismatch. Cracks perpendicular to the electrode edge develop due to the latter mismatch. This effect has been thoroughly discussed in Lucato et al. (2001b,a).

A different type of cracks was observed in cyclically fatigued circular discs of PIC 151. These are cracks parallel to the electrode faces underneath the electrodes as depicted in Fig. 3.12 Wang et al. (1998); Lupascu et al. (2000). Their origin is not exactly clear, but two effects may yield this crack geometry. First the perpendicular edge cracks discussed in the previous section may form the starter cracks for delamination. Once initiated, the modified stress field may deflect them to propagate underneath the electrodes. A second effect of crack initiation in the near electrode regions are the electrodes themselves. As Jiang et al. (1994b,a) pointed out, rough electrodes will induce field singularities at the sample surface. These in turn induce a significant strain mismatch between neighboring grains and immediately generate microcracks. The perpetual switching during fatigue then yields the driving force for fatigue crack growth. While in the case of cracks marked in the left image of Fig. 3.12 the tensile stresses necessary for crack propagation were easily determined (Lupascu et al., 2000), the K-criterion (stress intensity factor) for the delamination cracks on the right hand side of Fig. 3.12 is not entirely clear.

4 Acoustic Emission and Barkhausen Pulses

4.1 Acoustic Emission Technique

4.1.1 Principle

The acoustic emission technique is considered as one of the non-destructive testing methods for materials and devices (Beattie, 1976; Dunegan and Hartman, 1981; Wadley and Scruby, 1981; Wadley et al., 1981; Beattie, 1983). Different from most other methods, no external fields, radiation or sound waves are applied as inherent part of the method. All acoustic signals occurring at the surface of the sample originate from its interior. On the other hand acoustic emissions are only generated, if the material is loaded by some external forces, be they mechanical, electrical, magnetic, or thermal. All detectable interior changes of the material have to occur abruptly to induce a pulse of acoustic waves. The frequencies may be very low, like when considering the earths crust as a large sample and moving internal faults generating earthquakes as the acoustic emission events (Aki and Richards, 1980), but in technical systems ultrasonic frequencies are always encountered. The internal effects leading to acoustic pulses are cracking (Wadley et al., 1981; Wadley and Scruby, 1983; Krell and Kirchhoff, 1985; Sklarczyk, 1992), microcracking (Wadley and Simmons, 1987; Subbarao, 1991), friction (Wadley and Simmons, 1987), martensitic phase transformation (Speich and Schwoeble, 1975; Clarke and Aurora, 1983; Simmons and Wadley, 1984; Grabec and Esmail, 1986) or dislocation motion (Hatano, 1977; Rouby and Fleischmann, 1978; Scruby et al., 1981; Heiple et al., 1981) in metals or structural ceramics. In ferroelectrics the discontinuous motion of domain walls (Mohamad et al., 1979, 1981, 1982; Morosova and Serdobolskaja, 1987; Zammit-Mangion and Saunders, 1984), electromagnetically induced pulses (Aburatani et al., 1998) as well as the discontinuous generation or annihilation of domains (Kalitenko et al., 1980; Dul'kin et al., 1992, 1993; Dul'kin, 1999) may further contribute to acoustic events emanating from the sample.

For sufficiently large sources, homogeneous material, high local resolution of the detectors, and good time resolution of the electronics a detailed analysis of the acoustic emission transients may yield information on the source type itself (Ohtsu, 1995), but sophisticated deconvolution techniques have to be employed (Burridge and Knopoff, 1964; Aki and Richards, 1980).

4.1.2 Instrumentation

The AE-detection device used in this work is a (excellent) commercial system of bandwidth 100-2000 kHz (AMS3, Vallen Systeme, Icking, Germany). All the experimental results were obtained using 150 kHz resonant microphones (Physical Acoustics, Princeton, NJ, USA; or Vallen Systeme, Icking, Germany). This choice is somewhat disputable, because the typical emission spectra of acoustic pulses emanating from a ceramic microstructure occur in the MHz to GHz-range. If the propagation of a microcrack occurs typically at or in the proximity of the speed of sound, the time scale of the event is about 1 nanosecond (grain size 5 μm, elastic constant $s_{11} = 8 \cdot 10^{-12}\,\mathrm{m}^2/\mathrm{N}$, density $\rho = 7900\,\mathrm{kg/m}^3$). As this value is so far above the bandwidth of the setup, the resonance frequency of any detector falling into the bandwidth of the electronics will be equally bad. The acoustic emission transients thus display the resonant vibration response of the elastic system to some microscopic source exciting it. The whole process can be considered like the hitting of a resonant bell with its clapper, where the impact of the clapper is hardly noticeable, but a high frequency system would be able to tell differences in between individual impacts. In the case of the ceramic samples in this work, the sample plus detector constitute the resonating system. As the sample is small (thickness 1 mm) with respect to the detector (thickness 15 mm) the resonant frequency of the entire setup differs little from the detectors resonance itself. The question of the effective resonator will be reconsidered for the single crystals in Sect. 4.3.1.

The setup then registers the electric amplitude-maximum A and the event energy W as the integrated square of the amplitude for all consecutive threshold crossings which do not cease to occur for longer than a preset time interval. The data set between the first and the last threshold crossing fulfilling this condition is termed a hit H. Furthermore the number of threshold crossings within the hit, the time of the first threshold crossing, the duration of the hit and the entire transient are registered by the system. The transients can be subsequently analyzed using Fourier transforms. As the analysis features of the system were sufficient for the purpose of this work, the transients were not exported and externally analyzed.

The units for amplitude are given on a decibel (dB) scale with respect to 1 μV. The energy unit corresponds to 1 e.u. = 10^{-18} J electrical at the preamplifier input. Plots show "e.u." as energy unit, because a precise calibration of the acoustic energy is difficult without the knowledge of the transfer function of the sample and detector. As all ceramic samples were of identical size, the transfer function was considered identical. In the case of fatigued samples the elastic constants may have changed considerably due to the fatigue, but this fact has to be neglected, because a calibration of this modification is not feasible. As the decay constant for all acoustic events detected from ceramic samples in this setup was found to be identical, the energy and maximum-amplitude may be equivalently used ($W = \alpha A_{max}^2$,

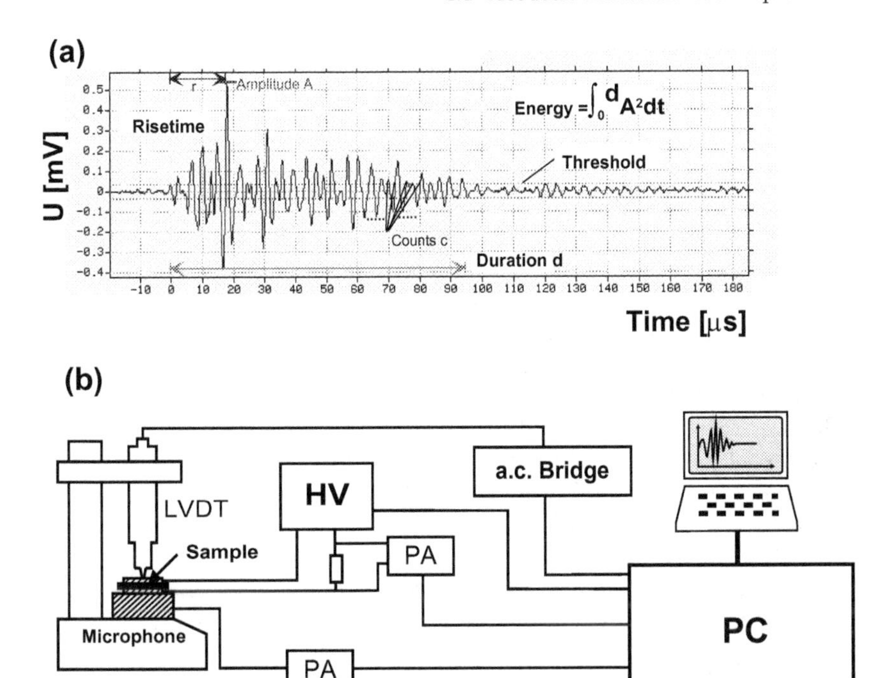

Fig. 4.1. (**a**) The transient of an acoustic emission displaying the maximum amplitude, threshold, threshold crossings, duration and time marker of an acoustic emission event. (**b**) Standard acoustic emission setup including the circuitry for the detection of Barkhausen pulses. PA = preamplifier, LVDT = linear variable displacement transducer, HV = high voltage source, PC = personal computer

with $\alpha = 6.8 \cdot 10^{-4}$ (e.u.)/(mV)^2). The voltage response detected for a single acoustic emission event in the acoustic emission setup are displayed on a typical transient recording in Fig. 4.1.

All ceramic samples were contacted by thin aluminum foil on the microphone side and a brass cone on the high voltage side providing electrical contact and reducing acoustic wave reflection. The samples were usually immersed in silicon oil to avoid arcing. The high voltage was generally applied at 20 mHz triangular bipolar voltage. In all setups a source of very narrow bandwidth near d.c. was used to reduce spurious electromagnetic noise (see Sect. 4.2.1). Acoustic coupling and electric insulation were provided by the silicon oil. In certain cases a resin for acoustic coupling was used.

4.2 Polycrystalline Lead-Zirconate-Titanate

4.2.1 Acoustic Emission Sources

Acoustic emissions from single crystals are limited to two processes, discontinuous domain wall motion as well as domain nucleation and annihilation. If a phase change were to occur, it would be directly visible on the outer habitus of the crystal as well as the unfortunate case that a crystal breaks. Nevertheless, the underlying causes for the discontinuous domain wall motion may be diverse. In polycrystalline ferroelectrics the realm of sources is much wider. Here microcracking (Kirchhoff et al., 1982; Dunegan and Hartman, 1981; Lupascu, 2001) can not easily be discerned from other local sources. Furthermore crystallographic phase changes may occur as well as dielectric breakdown in pores. The domain wall motion furthermore becomes a somewhat collective process and the origins of switching discontinuities may be different from the single crystal case.

Switching in ferroelectrics includes all processes that will change the externally observable polarization under the application of an electric field and thus the general term switching can not be appropriately used for denoting a particular acoustic emission source. The smooth motion of domain walls will not contribute to the acoustic emissions, while all discontinuous currents, namely ferroelectric Barkhausen pulses (Chynoweth, 1958), are likely to be accompanied by acoustic events (Zammit-Mangion and Saunders, 1984; Serdobolskaja and Morozova, 1998). It will be shown in Sect. 4.3.1, that the latter assumption is not generally the case, though. In all cases electromagnetically induced current pulses in the sample will via the piezoelectric effect induce an acoustic signal (see Table 4.2 and the discussion in Sect. 4.2.2).

During a phase transition the generation or on heating the annihilation of new domain walls is dominating the acoustic emission pattern (Dul'kin et al., 1993). The formation of new domain walls predominantly occurs at defects in the crystal, where stresses or local fields are enhanced. Also crack formation during the phase transition and crack healing for heating were assumed to be acoustic emission sources (Srikanth and Subbarao, 1992), but particularly the latter micromechanism seems unreasonable after the thorough analysis by Dul'kin et al. (1993).

Several studies have so far investigated the acoustic emissions under bipolar electric driving conditions (Arai et al., 1990; Saito and Hori, 1994; Uchino and Aburatani, 2000). Most acoustic emission events occur above the coercive field and very few under relaxation. A first source assignment has been attempted for polycrystalline materials (Lupascu and Hammer, 2002) and will be outlined in the following sections. Unipolar driving was shown to yield significantly less acoustic emissions than bipolar switching (Arai et al., 1990).

Ferroelectrics, when poled, are macroscopically piezoelectric. Thus every polarization change will, via the piezoelectric effect, induce a proportional stress (and strain) change at the surface of the sample, which can be misinterpreted as an acoustic emission. For a source volume on the scale of the microstructure, namely a volume of a grain or a pore, an estimate shows that the piezoelectrically induced acoustic emissions are five orders of magnitude smaller than the approximations given in Sect. 4.2.2. Thus piezoelectrically induced charge changes originating from the sample itself can be neglected. Nevertheless, it is evident, that external electromagnetic noise has to be well avoided in the setup, because these pulses may yield piezoelectric pulses of the same amplitude as true acoustic emissions from the sample (Aburatani and Uchino, 1996).

An illustration of the different acoustic emission sources can be found in Sect. 4.7 along with the discussion of the different cases occurring in polycrystalline PZT (p. 94).

4.2.2 Different Crystal Structures of PZT

Previous work showed, that differences in the c/a ratio will induce strong changes in the acoustic emission patterns in tetragonal $PbTiO_3$ (Choi and Choi, 1997). Thus differences in crystal structure for PZT are very much likely to generate different acoustic emission patterns. This section illustrates the different acoustic emission patterns generated by the tetragonal, rhombohedral and morphotropic PZT. A source assignment is attempted on the basis of different model approaches evaluating the acoustic emission output for microcracking, discontinuous switching, crystallographic phase changes and piezoelectrically induced events.

Acoustic Emission Data

The three lanthanum doped compositions as described in Sect. 1.5.3 were used. For all data, the amplification in the setup was 54 dB, the threshold 21.7 dB (= 12 mV). Two levels of maximum field were applied to the samples, 1.7 E_c and 4.6 kV/mm, the latter being about 2 E_c for the tetragonal material (highest E_c). In Fig. 4.2 the acoustic emission amplitudes during one hysteresis cycle are plotted versus time for both maximum fields in the coarse grained compositions, taken a few cycles after the first poling. Each event is represented by a dot. For $E = 1.7\ E_c$ the amount of acoustic emissions is highest for the tetragonal composition and lowest for the rhombohedral composition. All amplitudes up to about 60 dB occur. In the morphotropic composition a gap of amplitudes is found between 50 and 70 dB. Amplitudes above 70 dB only occur in the morphotropic composition. The rhombohedral composition hardly shows any events at all.

If the identical maximum field (4.6 kV/mm) is applied to all compositions, the most pronounced occurrence of acoustic emissions is observed in the morphotropic material (right column in Fig. 4.2). The rhombohedral composition

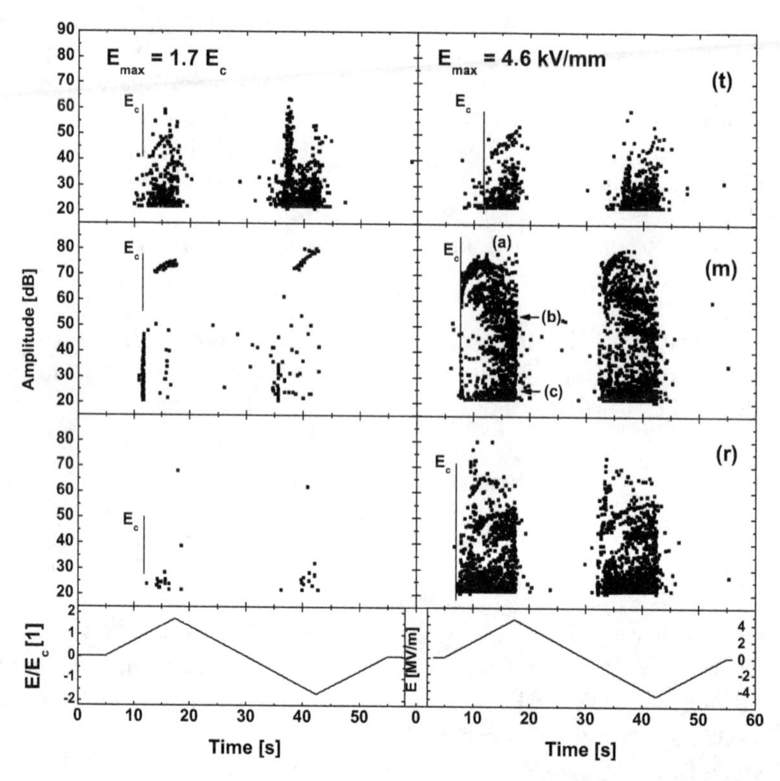

Fig. 4.2. Time dependent AE-patterns for coarse grained (t) tetragonal, (m) morphotropic and (r) rhombohedral PZT (2% La)

Table 4.1. Cumulative AE - hits and AE - energy during 5 consecutive hysteresis cycles after the first bipolar cycle. * The values for the coarse grained morphotropic material are determined after dynamic relaxation

		$E_{max} = 1.7\,E_c$		$E_{max} = 4.6\,\mathrm{kV/mm}$	
Structure	Grain size	Hits	Energy [10^4 e.u.]	Hits	Energy [10^4 e.u.]
tetragonal:	fine	17000	14000	17000	14000
	coarse	8000	700	8000	700
morphotropic:	fine	1200	43	21000	45000
	coarse*	10500	4300	29000	38000
rhombohedral:	fine	750	6	15000	79000
	coarse	220	250	26000	63000

also shows enhanced acoustic activity. A range of many low amplitude events is found in all materials (c).

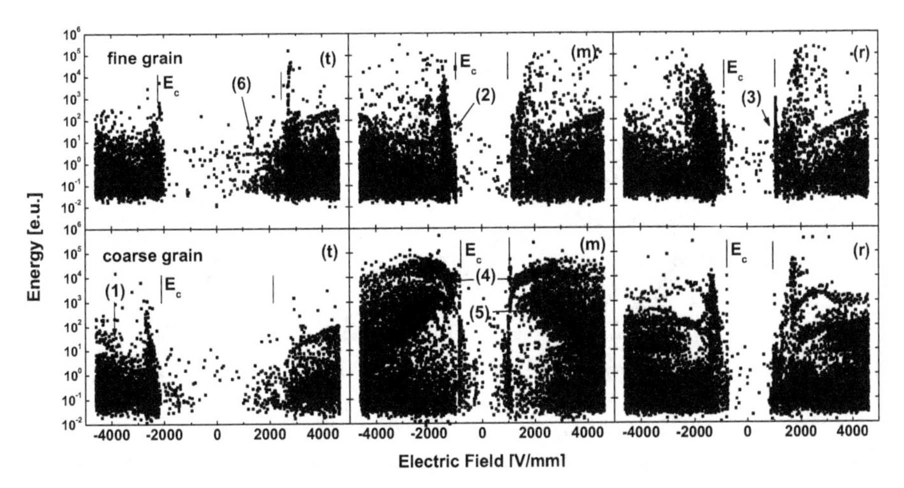

Fig. 4.3. Acoustic emission energies vs. electric field for tetragonal (t), morphotropic (m) and rhombohedral (r) PLZT averaged over 5 cycles, E_{max} = 4.6 kV/mm. The numbers indicate different types of acoustic emissions

In the morphotropic composition, the few events found at high amplitudes for $E = 1.7E_c$ now form continuous streaks to higher amplitudes at higher fields (event types (4) and (5) in Fig. 4.3). At higher field levels events of medium amplitude become more numerous. Table 4.1 summarizes the differences of total energy dissipated during five cycles for both voltages.

Essentially all acoustic emissions occur above the coercive fields. For decreasing fields the acoustic emissions immediately cease in all materials.

Poling and the first inversion cycle are different. Not only that the first poling will yield stronger acoustic emissions than the subsequent cycles but some acoustic emissions will also occur below the coercive field during the first poling inversion irrespective of the crystal phase. The ranges of acoustic emission occurrence are also marked in Fig. 1.6.

Figure 4.3 displays the field dependent acoustic emission patterns for all 6 materials. Common to all materials are a base of low energy events at high fields (1) (events (c) in Fig. 4.2) and high energy events just above the coercive field (2). For the morphotropic and rhombohedral composition further events occur in a very narrow field range at E_c (3), Fig. 4.3. They are distributed across all strain values which are passed at E_c. In the coarse grained morphotropic material streaks of high energy events (4) analogous to Fig. 4.2 (a) and events of medium energy (5) ((b) in Fig. 4.2) are found. Slight indications of such events also occur in the coarse grained rhombohedral composition.

In the tetragonal materials a few events occur beneath E_c for rising fields during the first cycle after some resting time (6) indicating a relaxation process in the domain structure due to local stress fields. This type of event does

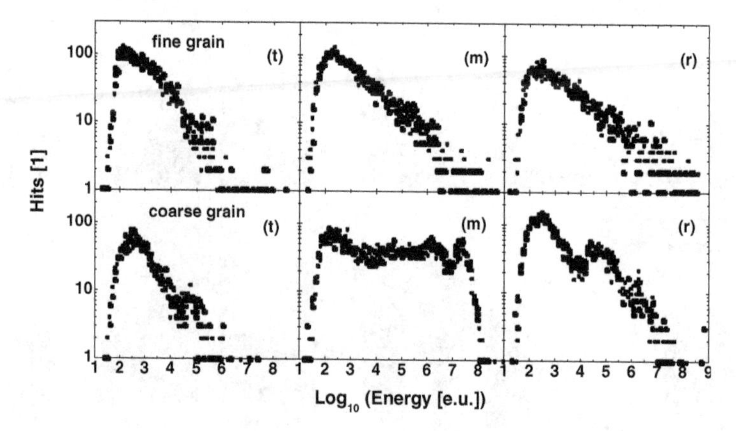

Fig. 4.4. Cumulative AE-energy distribution for 5 consecutive cycles for tetragonal (t), morphotropic (m) and rhombohedral (r) coarse and fine grained PLZT, E_{max} = 4.6 kV/mm

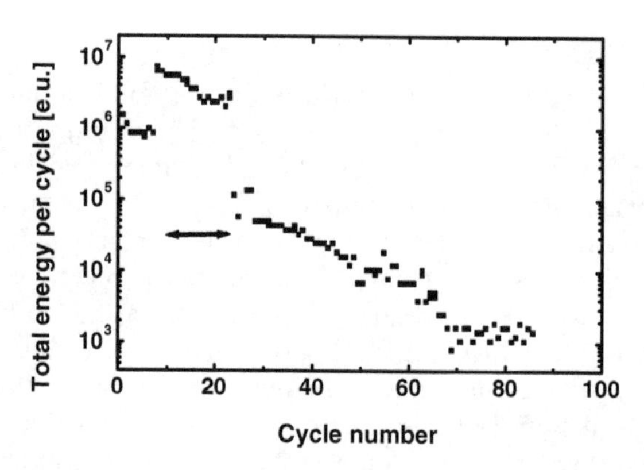

Fig. 4.5. Total dissipated energy per cycle as a function of cycle number (E_{max} = 1.8 kV/mm). The arrow indicates 12 intermediate cycles, when the material was driven up to 4.6 kV/mm

neither occur for the subsequent polarization inversion nor in the immediately following cycles. The material thus needs a certain time to relax before showing this type of event.

The energy distributions of the acoustic emission events from Fig. 4.3 are shown in Fig. 4.4. The fine grained rhombohedral and morphotropic materials show a power law dependence of the number of hits H with respect to their energy: ($H = \beta E^{\alpha}$). The fine grained tetragonal material displays some deviation from a straight line (exponents α = -0.28 (r), α = -0.33 (m) and α = -0.55 (t) for high energies). In all spectra, the left edge of the spectrum

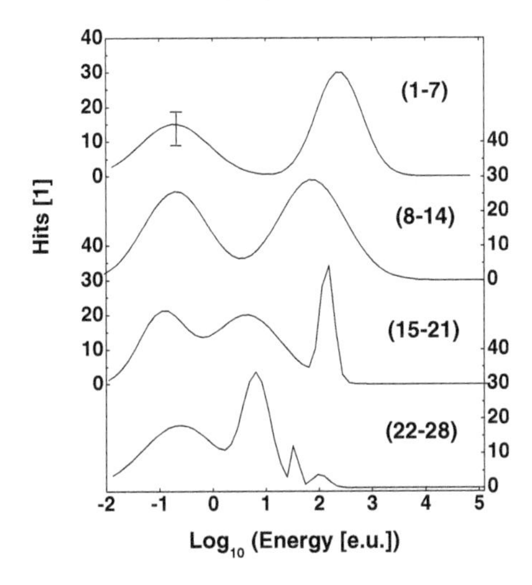

Fig. 4.6. AE-energy distribution (average over 7 cycles) for the time dependence of the acoustic emissions in coarse grained morphotropic PLZT. The cumulative distributions are approximated by lines for better visibility. An average scatter bar is shown

is due to the finite amplitude threshold set by the equipment. The coarse grained compositions, on the other hand, show significant amounts of high energy events which seem to group in distinct energy ranges. They are far from showing a power law relation.

Like the macroscopic data show relaxational behavior, also the acoustic emission pattern in the coarse grained morphotropic composition dynamically "relaxes" approximately during the first 80 cycles. This behavior is shown in Fig. 4.5 and 4.6. Figure 4.5 shows the energy integrated during full hysteresis cycles for each of the first 80 cycles versus cycle number. A whole spectrum like the ones shown in Fig. 4.2 is integrated to yield one total energy. 10 intermittent cycles were driven to 4.6 kV/mm, while all others were driven to 1.7 E_c. The significant reduction in energy is obvious. Figure 4.6 displays the same data as shifts in energy distribution within each cycle (averaged over 7 cycles for better statistics, taken after the high field cycles). The high energy events significantly decrease in number. The data taken at 4.6 kV/mm show an order of magnitude higher total acoustic emission energy.

Irrespective of the bipolar field values, the total energy decreases with each cycle by about the same relative decrement on a logarithmic scale. Finally, the total energy reaches values similar to those of the other compositions after about 90 cycles. Only events of type (1) and (2) of Fig. 4.3 occur at this stage.

Energy and Amplitude Estimates

The total energy liberated by one single grain of average grain size is evaluated for microcracking, domain switching, crystal phase changes and piezoelectrically induced pulses using different models for acoustic emission output. The different microscopic sources are compared for their effective acoustic emission amplitudes given as volume fraction of a grain sufficient to yield acoustic emissions above the experimental threshold.

Total Energy

Considering a crack tip toughness of PZT of $K_{Io} = 0.5\,\mathrm{MPa}$ (Chen et al., 2001), an average Young's modulus of $Y = 104\,\mathrm{GPa}$ (t), $Y = 85\,\mathrm{GPa}$ (m), $Y \approx 84\,\mathrm{GPa}$ (r) (Pferner et al., 1999) and Poisson's ratio of $\nu = 0.35$ (Jaffe et al., 1971), neglecting the anisotropy of Young's modulus and the domain structure within the grains, the mechanical energy release rate for crack advance (Lawn, 1993) becomes G $= 2.1\,\mathrm{J/m^2}$ (t); $2.6\,\mathrm{J/m^2}$ (m) and $2.6\,\mathrm{J/m^2}$ (r). Considering transgranular cracking and spherical grains for simplicity, the fracture energy for one grain will be $5.4 \cdot 10^{-12}\,\mathrm{J}$ (t); $6.6 \cdot 10^{-12}\,\mathrm{J}$ (m) and $6.6 \cdot 10^{-12}\,\mathrm{J}$ (r) for the fine grained material ($1.8\,\mu\mathrm{m}$), or $1.5 \cdot 10^{-11}\,\mathrm{J}$ (t); $1.8 \cdot 10^{-11}\,\mathrm{J}$ (m) and $1.8 \cdot 10^{-11}\,\mathrm{J}$ (r) for the coarse grained ($3\,\mu\mathrm{m}$) material. Higher energies will be constituted by instantaneous crack formation larger than one grain. The energies given will be dissipated into the formation of new surface area and the generation of acoustic waves. Thus, only a fraction of the energy will be detected as an acoustic wave.

To estimate the total electric switching energy for instantaneously switching an entire grain, the macroscopic hysteresis data are used. Considering an average polarization of $0.4\,\mathrm{C/m^2}$ and a coercive field of $1\,\mathrm{kV/mm}$ (in the tetragonal material the coercive field is twice as high, but the polarization half as much), the energy density is $8 \cdot 10^5\,\mathrm{J/m^3}$ (full reversal is twice the remnant polarization). The total energy needed to switch one grain then becomes $2.4 \cdot 10^{-12}\,\mathrm{J}$ (f.g.) or $1.1 \cdot 10^{-11}\,\mathrm{J}$ (c.g.). Essentially the energy needed for cracking one grain is in the same order of magnitude as for switching one grain of the same size. As the grain size distribution contains some grains of sizes up to $7\,\mu\mathrm{m}$ (Hoffmann et al., 2001), events of considerably higher energy will be encountered. Jiang et al. (1994e) assigned acoustic emission events of several orders of magnitude difference to microcracking and domain wall motion. Such large differences between both acoustic emission origins seem unreasonable in the light of this first energy estimate.

Acoustic Emission Amplitudes

Amplitude estimates are utilized here to assign the different acoustic emission sources. Two theoretical approaches are known, which convolute the bandwidth of the detector with the frequency spectrum of the microevent. Malén and Bolin (1974) assume a Heaviside function as the source function for the strain change s_{ij} of a representative acoustic emission source volume ΔV. The transfer function of the detector is given by the term $\omega\Delta\omega$, with the detector bandwidth $\Delta\omega$ and the resonant frequency ω. For evaluating the stress amplitude:

$$\sigma = \frac{\omega\Delta\omega\rho}{2\pi^2 r}\sum s_{ij}\Delta V \tag{4.1}$$

the material density ρ and the distance r between detector and source also have to be known. The factor 2 in the denominator is valid for pure shear and has to be replaced by a factor 4 for hydrostatic volume changes. Simmons and Wadley (1984) arrive at a similar expression for the stress amplitude being proportional to the strain times the active volume generating acoustic emissions for martensitic transformations, but they do not specify the transfer function of the setup. For discontinuous domain wall movement we consider a tetragonal unit cell to be switching by 90° through its orthogonal state thus yielding a brief volume change in field-direction of $\Delta s_{33}\Delta V = (c/a - 1)\, V_{grain}$, where c and a are the lattice constants along the long and short tetragonal axes and ΔV the volume of the grain. For microcracking Malén and Bolin (1974) used an Eshelby-technique. The amplitudes for propagation of an infinite slit-like crack are evaluated according to:

$$\sigma = \frac{\omega\Delta\omega}{4\pi^2 rc^2}\sigma_o h\left(l_2^2 - l_1^2\right) \tag{4.2}$$

where σ_o is the external stress present at the moment of cracking, h the finite height in the direction of "infinite" slit length, l_1 and l_2 the half lengths of the crack before and after crack advance in the crack propagation direction and c the longitudinal speed of sound (isotropic medium approximation). The stress amplitude σ in either case translates into an electrical amplitude A via $A = g_{33} l_{det}\sigma$ with the piezoelectric stress constant g_{33} and the thickness l_{det} of the detector.

Wadley and Simmons (1987) determined the response of a detector with Gaussian frequency distributions of bandwidth B and to a source of duration τ to be:

$$\kappa(\tau) = \frac{\tau B}{\sqrt{1 + \tau^2 B^2}} \tag{4.3}$$

which is effectively τB for the grain sizes and bandwidth in our experiment. They arrive at an expression for the cracking induced displacement u at the sensor surface including the transfer function κ and the crack propagation speed $\dot{l}$:

$$u_3 = \left[\frac{4(2-\nu)(1-\nu)^2}{\pi\rho(1-2\nu)c^3}\right]\frac{\kappa(\tau)\sigma_3 l^2 \dot{l}}{r_3}, \tag{4.4}$$

where σ_3 and r_3 are equivalent to σ_o and r of equation 4.2 and the index 3 refers to the polarization direction. The electric voltages are determined from $A = u_3/d_{33}$ with the piezoelectric constant d_{33} of the detector. Equations 4.1 and 4.2 will be compared in order to estimate differences between

Table 4.2. Acoustic emission amplitudes calculated according to the referenced equations. Given is the number of grains necessary to constitute an acoustic emission event for threshold crossing (assumed detector-distance from the acoustic emission source location 1.5 mm, grain sizes see Sect. 1.5.3 and Table 1.3)

Material (Equation)		Switching (4.1)	μ-cracking (4.2)	μ-cracking (4.4)
tetragonal:	fine grain	6	70	2
	coarse grain	1	15	0.4
	max	0.1	1.8	0.04
morphotropic:	fine grain	5	37	1.9
	coarse grain	1.5	13	0.5
	max	0.1	1.4	0.05
rhombohedral:	fine grain	20	42	1
	coarse grain	3	9	0.2
	max	0.3	1.4	0.03

polarization switching and cracking and 4.2 and 4.4 for estimating the influence of the bandwidth of the detector and the finite times and speeds of the sources. Application of equations 4.2 and 4.4 requires knowledge of the microcracking stress σ_o. Unfortunately, the microcracking criterion is still subject to discussion (Zimmermann et al., 2001; Vedula et al., 2001). The size of the critical nucleus, which pops to the full grain facet length, is the most uncertain parameter. A typical assumption is, that a nucleus at a triple point is viewed as a critical defect over which stress can be integrated to arrive at a stress intensity factor. This can then be compared to the grain boundary toughness. Typically a circular crack 2 l_{start} of half the grain size g: $l_{start} = g/4$ is assumed (Lawn, 1993), because a strong influence of grain size on microcracking has been observed (Paulik et al., 1996). I use:

$$\sigma_o = \frac{\sqrt{\pi}}{2} \cdot \frac{K_{Ic}}{\sqrt{l_{start}}} \tag{4.5}$$

which is a function of grain size accordingly and assumes the slit like cracks (Malén and Bolin, 1974) to be of identical area as the circular starter crack: $hl_2 = \pi l_2^2$. The effectively applied K rises strongly for increasing crack length in the case of microcracking and entire grains are likely to crack in a single event. I thus assume the crack propagation $\dot{l}$ to occur at the speed of sound c. The dielectric and elastic constants are in each case identical for microcracking and domain switching and will in all cases be approximated by the isotropic values of an unpoled sample. The most uncertain parameter in these approximations is the detector distance r, which unfortunately is small with respect to the sample and detector dimensions. As it only enters as a scaling factor $1/r$ in all three equations 4.1, 4.2 and 4.4, the absolute energy values may differ considerably, while their relative difference will remain identical. An effective distance to the detector surface of 1.5 mm was chosen.

Numerical Values

Values for the largest grains are assumed, because they are those that are most likely to exceed the threshold value and detected during experiment (7 μm in all compositions). The density is $\rho = 7900\,\mathrm{kg/m^3}$, the elastic moduli were given on p. 90 and the equivalent cubic lattice distortions $(c/a-1)$ (Hoffmann et al., 2001) are: 0.0276 {0.0277 (c.g.)}(t); 0.0200 {0.0201(c.g.)}(m) and 0.0062 {0.0065(c.g.)}(r). The tilt angle in the rhombohedral compositions is not included. The detector parameters are $g_{33} = 2.5 \cdot 10^{-3}\,\mathrm{Vm/N}$, ν=150 kHz, $\Delta\omega = 50\,\mathrm{kHz}$, and $l_{det} = 15\,\mathrm{mm}$. Table 4.2 summarizes the estimated voltage amplitudes given as the equivalent number of grains necessary for threshold crossing (12 mV).

Discussion

The discussion is divided into two parts. First the field ranges, strain ranges, and external driving conditions yielding acoustic emission are discussed. Then a source assignment is attempted.

Driving Conditions for Acoustic Emission Generation

Almost all acoustic emissions occur for rising fields at or above the coercive field as marked in the hysteresis loops in Fig. 1.6. Similar ranges were found by Mohamad et al. (1979) in $Pb_5Ge_3O_{11}$ and by Uchino and Aburatani (2000) in PZT, but implications on the generation mechanism were not discussed there. The acoustic emissions arise in the saturation range of the dielectric as well as strain hysteresis with most events occurring right above E_c. Most polarization changes have already taken place. The mere switching therefore is not a sufficient source of acoustic emission. As most 180° switching has taken place above E_c, the acoustic emissions are predominantly due to 90° switching, but 180° switching can not be entirely excluded like the single crystal data in Sect. 4.3.1 demonstrate. High external driving forces are needed for acoustic emissions to occur. The driving forces present during relaxation are not sufficient to induce discontinuities in the domain switching process. The underlying micromechanism may be viewed similar in character as the Orowan process in dislocation motion (Hull and Bacon, 1984), where dislocations are clamped at some inclusion and discontinuously released (Dunegan and Hartman, 1981). Evidently, a domain wall is a planar defect, and the dislocation a linear one, but the principle is similar. The domain wall gets stuck behind an obstacle like the point defect agglomerates discussed in Sects. 3.1 and 5.8.

During first poling and the first poling inversion, acoustic emissions were also observed beneath the inflexion points of the polarization and strain hysteresis irrespective of crystal structure. As the material starts off in a completely disordered state after sintering, even low fields will be sufficient to

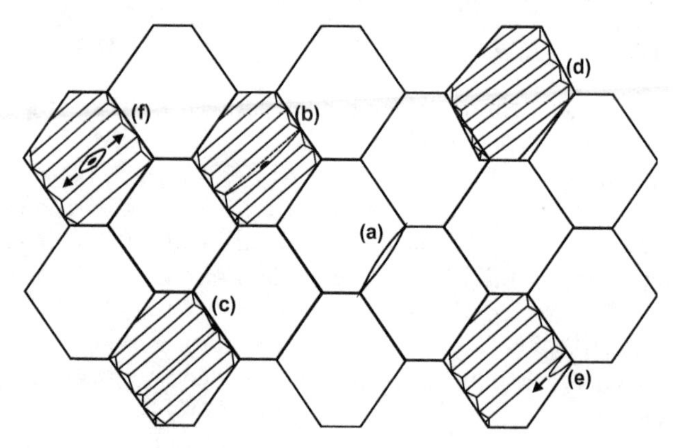

Fig. 4.7. Different acoustic emission sources in the microstructure of a ferroelectric ceramic. (**a**) Microcracking, discontinuous jumps of a domain wall clamped at an internal defect (**b**) or a defect at the grain boundary (**c**), (**d**) crystallographic phase change of entire grains, (**e**) domain wall generation at a grain boundary, (**f**) domain wall generation at a defect. Illustrated is a cut through a filling of space by icosahedral grains of equal grain size

discontinuously switch some domain systems across microscopic obstacles or induce a crystal phase change. Furthermore, local stresses due to cooling may still be present and represent an additional driving force, no longer available in subsequent cycles. All types of the above mentioned events are likely to occur.

The tetragonal composition is the only one that shows a few acoustic emissions for decreasing fields. Some events are also always found beneath E_c for the first cycle after a rest time. Values of E_c in tetragonal PZT are strongly rate dependent. If the coarse grained material of our study is allowed 2 hours for one cycle, E_c drops down to about 1.25 kV/mm (Hoffmann et al., 2001), which is almost 50% of the values determined here (50 seconds). The material will thus creep. As the sample is not given enough time to relax completely, events which in the other compositions would have occurred immediately, arise later when the voltage is still higher than the threshold for this particular obstacle, but already decreasing overall. Similarly the material will creep back into a stable configuration, if given enough time.

Assignment of Acoustic Emission Sources

The energy estimates from Sect. 4.2.2 showed that large grains constitute sufficiently large volumes to generate detectable acoustic emissions irrespective of the source type and the particular equations used for the evaluation. Thus, it is reasonable to discuss sources on the scale of the microstructure.

Piezoelectrically induced events can be neglected according to Sect. 4.2.1. An easily assignable source is the crystallographic phase switching in the coarse grained morphotropic composition. This type of extremely large and numerous events ((4) and (5) in Fig. 4.3) only occurs for this composition and totally ceases after about 90 cycles, while the acoustic emission patterns in the other compositions essentially remain unchanged. The unit cell volume change during the morphotropic phase transformation ($\Delta V/V = 3.4\%$, Hoffmann et al. (2001); Xu (1991)) is 1.7 times larger than for pure switching. It is still deviatoric in character and the factor 2 in the denominator of (4.2) remains valid. This type of source will therefore generate higher energies than switching and microcracking. The state of best orientation of the grains with respect to the externally applied field is approached, until the entire microstructure has best adopted to the texture imposed. This mechanism is similar to acoustic emissions occurring at the phase transition from cubic to tetragonal at the Curie point known for $BaTiO_3$ (Dul'kin et al., 1993) or acoustic emissions due to phase switching in zirconia (Clarke and Aurora, 1983).

Microcracking occurs at high field values and yields moderate acoustic emission amplitudes. Purely electrostrictive PLZT also generates acoustic emissions yielding a power law amplitude distribution (Lupascu, 2001). This material contains no domains and only microcracking can (Haertling, 1987), and was shown to occur (Lupascu et al. (2001), see Sect. 3.7.1). A ferroelectric composition of identical Zr/Ti ratio showed additional events associated with domain processes (Lupascu, 2001). Within the same model (equations 4.1 and 4.2) discontinuous switching will yield higher amplitudes than microcracking. Events of type (1) rather than type (2) in Fig. 4.3 are therefore assigned to microcracking. At high field values all ferroelastic switching has ceased. All grains are fully oriented. The electric field then generates high piezoelectric strains, which directly translate to stresses. Domain wall motion can no longer accommodate these strain changes and the stress values become sufficiently large to initiate microcracking.

Discontinuous domain switching occurs in two forms. The rhombohedral and to some extent the morphotropic compositions yield acoustic emissions right at E_c (type (3) in Fig. 4.3), when the largest and fastest macroscopic strain changes occur. For slow strain changes smaller obstacles in the path of the domain wall propagation are usually circumvented by the domain system and slowly overrun. If the external driving forces, in this case the strain change of the entire neighboring sample, change rapidly, the obstacle is overrun at a sufficiently high speed to generate detectable polarization and strain jumps. This type of event can not occur in tetragonal compositions, because the switching of the entire sample occurs slowly and the coercive field is smeared out. The second type of discontinuous domain switching occurs right above E_c (events (2) in Fig. 4.3). It is due to larger obstacles clamping the domain system in a particular grain. When neighboring grains have fully aligned with the external driving force, the stresses/fields acting on the clamped domain system become sufficiently large to overrun the energy barrier and the domain system releases its energy.

To some extent streak-forming events similar to events of type (5) in Fig. 4.3 can be found for the large grained rhombohedral composition also forming extra peaks in the energy distribution (Fig. 4.4). An assignment of these events is not evident.

Summary

Acoustic emissions only occur during particular parts of the hysteresis loop, when the domain system is driven into regimes, where the local constraints become severe. The local discontinuities may then be domain nucleation and annihilation at grain boundaries or defects, the discontinuous jumps of single walls, or, if the stresses become even larger, microcracking. As will be shown in Sect. 4.2.4 the application of a uniaxial compressive stress along the field direction can considerably reduce these effects, even though the polarization hysteresis saturates.

The poling cycle in all PZT compositions shows a different acoustic emission pattern. This is intuitively understandable, because severe rearrangement of the domain system by 90° switching has to occur, part of which has to surmount fairly high energy barriers. This is not the case in the subsequent cycles, when, even though 90° switching occurs, it effectively yields a 180° polarization inversion for the entire cycle.

To a fair degree of confidence, acoustic emission events due to crystallographic phase switching, polarization switching and microcracking were separated for the three compositions of soft 2%La-doped PZT. Sources on the scale of single grains are calculated to be sufficiently large to yield detectable acoustic emissions. An induced crystallographic phase change during a first few cycles was found for some grains in the morphotropic material, while some relaxation was observed in the tetragonal modification.

The processes inducing acoustic emissions and those responsible for fatigue are likely to be closely related. The range of fields for considerable acoustic emissions is just above the coercive field, which is the same range of fields that has to be reached to induce significant fatigue of PZT.

4.2.3 Fatigue Induced Discontinuities

Due to bipolar fatigue the domain wall movement is significantly altered in PIC 151 (Nuffer et al., 2000). This is an effect that was reproducible in every bipolar fatigue measurement. The absolute energies rise by a few orders of magnitude indicating enormous changes in the microscopic structure of the ferroelectric ceramic. As was already depicted in Fig. 2.15 unipolar fatigue yields a significantly different behavior. After fatigue, not an increase of the AE activity is seen, but a significant decrease. This is another crucial observation.

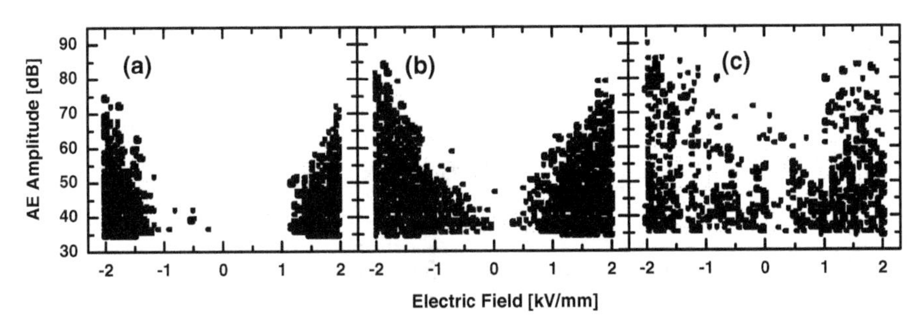

Fig. 4.8. Acoustic emission patterns from PIC 151 at different fatigue stages (1.96 kV/mm, 50 Hz), (**a**) unfatigued, (**b**) $3 \cdot 10^6$ cycles, (**c**) 10^8 cycles

The unipolar fatigue changes the microstructure into a state, where domain wall motion encounters less obstacles, while after bipolar fatigue an enormous increase of energies needed to surpass obstacles is observed.

Figure 4.9 displays the detailed results of an acoustic emission measurement before and after bipolar fatigue. It is clear from the hysteresis data (a) and (b) that a severely damaged sample is encountered on the right hand side. Acoustic emissions only occur, when the external driving forces are large. This is particularly the case above the coercive field in the unfatigued sample. In the fatigued sample acoustic emissions already occur beneath the coercive field as determined for the unfatigued sample (f) and (g). Once the external driving forces decrease in value, the acoustic emissions as well as the Barkhausen pulses cease (d) and (e). This is the case irrespective of whether the sample was fatigued or not. In an unfatigued sample the acoustic emissions and the Barkhausen pulses both only occur at high polarizations, i.e. when the material is driven far into saturation (h), similar to all the material behavior observed in Sect. 4.2.2. In a fatigued sample the situation is entirely different. Here, every change in polarization is accompanied by significant acoustic emissions and Barkhausen pulses.

Another interesting effect is the high degree of coincidence between acoustic emissions and Barkhausen pulses (j). Most events coincide within a few microseconds and are thus due to the identical microevent. Altogether only a few more Barkhausen-events occur, but this absolute comparison is only vague, due to the inherently necessary introduction of a threshold during measurement.

4.2.4 Acoustic Emissions under Uniaxial Stress

Another interesting facet of acoustic emissions generated in polycrystalline materials is their reduced occurrence, if an external uniaxial compressive stress is applied (see Sect. 2.1). For pressures up to -28 MPa the ferroelectric

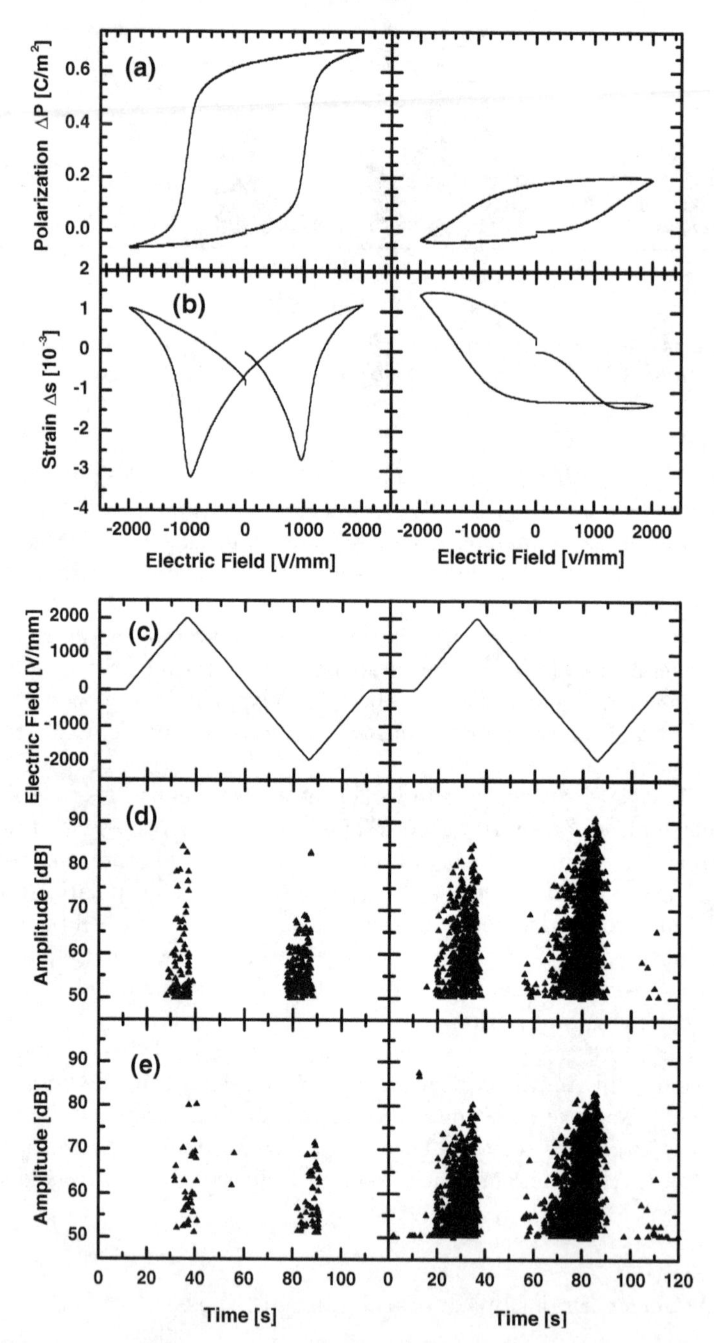

Fig. 4.9. Acoustic emission (AE) and Barkhausen (BHP) patterns from PIC 151 (0.5 mm thick sample) at 1 cycle (left) and 10^7 cycles (right) after bipolar fatigue ($2\,E_c$). (**a**) Polarization hysteresis, (**b**) strain hysteresis, (**c**) electric field vs. time, (**d**) AE and (**e**) BHP at field values from (**c**)

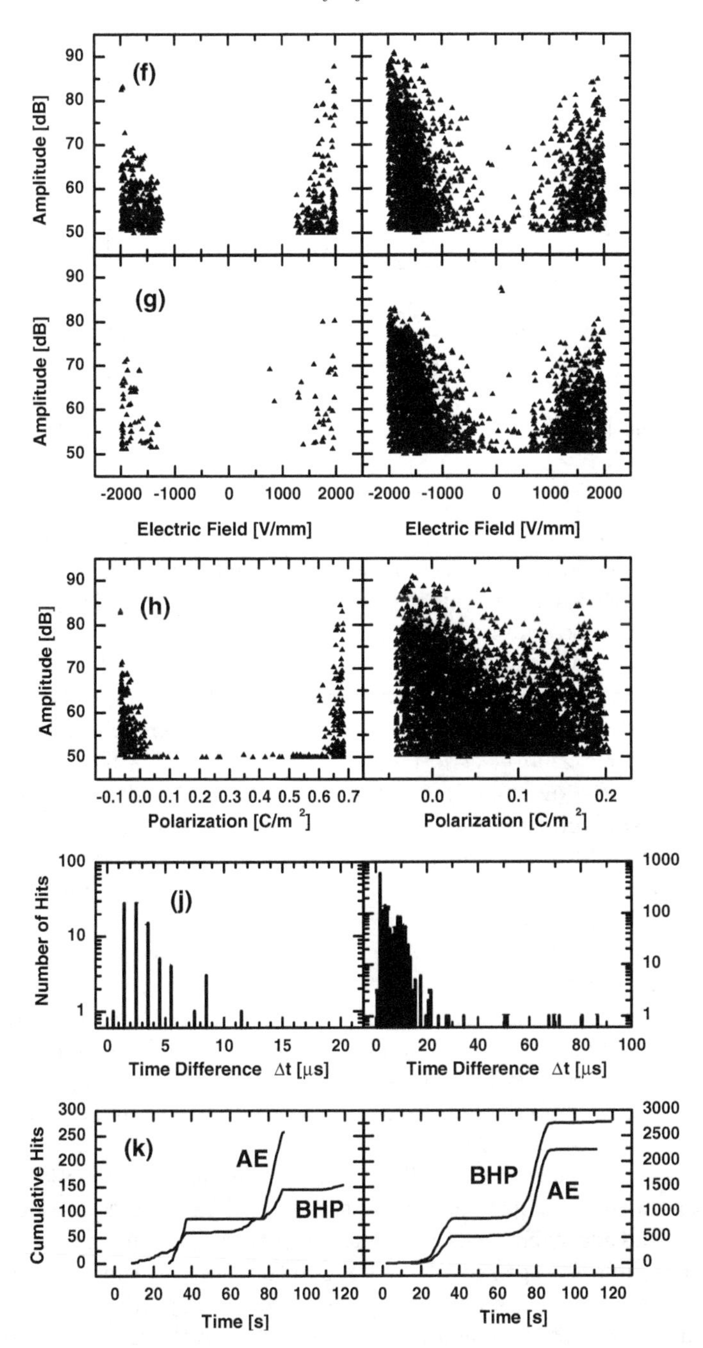

Fig. 4.9. Continued. AE (**f**) and BHP (**g**) as a function of electric field, all events as a function of polarization (**h**), (**j**) number of coincident hits and (**k**) cumulative number of hits for AE and BHP

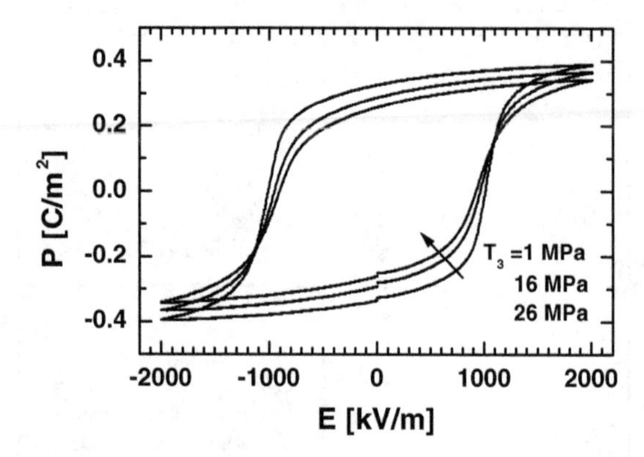

Fig. 4.10. The ferroelectric hysteresis of PIC 151 for different applied uniaxial stresses

hysteresis will only be decreased by about 10%, while the acoustic emission energy generated during one hysteresis cycle is reduced to 20% of its initial value. The coercive stress in this material is about -40 MPa. Thus the highest externally applied stress (-28 MPa) is about 70% of the coercive stress value. As it is applied along the same axis it essentially reduces the effect of the electric field. As the 180° domain walls are insensible to stress, the contribution of 90° switching to the ferroelectric hysteresis is reduced by this stress. The maximum external electric field exceeds the coercive field by a factor of 2. Therefore, the overall effect of the stress will be to reduce the effective electric field to a value just slightly above the coercive field. Actually much lower stresses (-7 MPa) are sufficient to reduce the acoustic emission energy to about 30% of its unclamped value. If only the macroscopic behavior were relevant, this compressive stress would not be sufficient to reduce the acoustic emission activity and emissions should equally well occur in great number. As this is not the case, the stress acts in a different fashion. It suppresses those parts of the domain wall motion, which are likely to be discontinuous. These contributions must be predominantly of 90° switching character, due to the strong influence of the stress. So, whatever obstacles impede the domain wall motion, they are no longer surpassed, once a slight compressive stress is opposing the field induced stresses.

These data correlate fairly well with the assignment of acoustic emission sources to the microscopic mechanisms discussed in Sect. 4.2.2. There the smallest amplitude events were associated with the occurrence of microcracking. The uniaxial stress will predominantly suppress the 90° switching, which due to constant volume (Poisson's ratio ν=0.5) will induce a transverse strain of 50% of the longitudinal strain. Due to the external stress this contribution is suppressed accordingly. The local stresses no longer exceed values necessary to induce microcracking. Already small external stress values are sufficient to reduce the number of small energy acoustic emissions

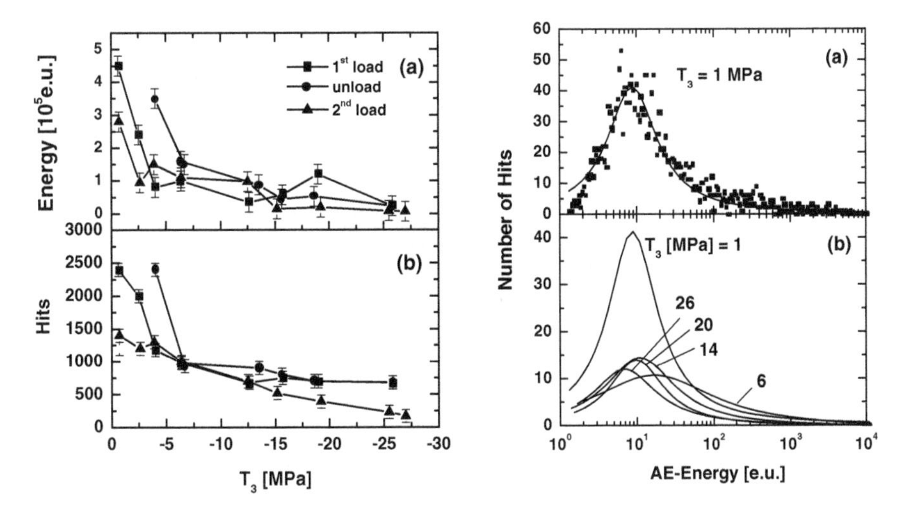

Fig. 4.11. (Left) Total number of hits and integrated acoustic emission energy for one hysteresis cycle at different applied stresses. (Right) The corresponding distributions of AE energies

drastically, while all other microscopic mechanisms still exist in the acoustic emission patterns, which is well visible in Fig. 4.11.

4.3 Single Crystal Ferroelectrics: Domain Wall Motion, Nucleation, and Annihilation

4.3.1 The Uniaxial Ferroelectric Gadolinium Molybdate

The effect of extended defects on the domain wall motion and the resulting acoustic emission (AE) and ferroelectric Barkhausen pulse (BHP) patterns in ferroelectrics are investigated in more detail in this section. For this purpose only domain switching in its most general sense is considered here. Crystallographic phase switching or microcracking are excluded.

The polarization reversal process itself is a complicated process of domain nucleation and subsequent domain growth by wall motion (Strukov and Levanyuk, 1998; Lines and Glass, 1977; Fatuzzo and Merz, 1967). The discontinuous parts of this process can be accessed by AE (Saito and Hori, 1994) and BHP (Lines and Glass, 1977; Fatuzzo and Merz, 1967; Chynoweth, 1959), if these are well interpreted. The observation of BHP and AE during the field induced motion of a single domain wall reduces the great variety of competitive BHP and AE sources in ceramics to the simplest process of domain wall kinetics being tackled now.

The well known improper ferroelectric-ferroelastic gadolinium molybdate $Gd_2(MoO_4)_3$ (GMO) is an ideal model material for the investigation of the

Fig. 4.12. View onto the GMO crystal along the orthorhombic c-axis [001] under crossed polarizers. The single planar domain wall is visible at its stable position. Regions of crystal defects left (A) and right (B)

sideways motion of a planar 180° domain wall (Kumada, 1969; Flippen, 1975; Shur et al., 1986, 1990). Previous investigations by Shur et al. (1999) displayed BHP for one such domain wall moving across an artificial electric field inhomogeneity near the edges of a GMO crystal under ac driving. Therefore, acoustic emissions were anticipated to also occur, when the force driving a domain wall over an extended defect exceeds a particular value and the wall is elastically released generating a discontinuous jump. Simultaneous to the measurement of the AE and BHP, the optically observable jumps of the domain wall were registered.

Orthorhombic $\beta' - Gd_3(MoO_4)_3$ (Keve et al., 1971) crystals were grown using the Czochralski method. Particular growth conditions were employed to introduce larger defects into the crystal (see Fig. 4.12, Bunkin and Nishnevitch (1995)). These are intended to stop the domain wall motion at certain positions in the crystal and may be called larger pinning centers. The crystal surfaces were oriented along the (001) and (110) planes. A single 180° domain wall normal to the (110) face separates two domains of point inverted symmetry (Kobayashi et al., 1972). In_2O_3:Sn-electrodes were sputtered onto the crystal in form of two U-shaped electrodes (see Fig. 4.13, Shur et al. (1999); Barkley et al. (1972)). A partial view along the [001] direction (crystal of dimensions 17.0 x 2.15 x 0.87 mm^3, edge lengths along $[1\bar{1}0], [110], [001]$) is shown in Fig. 4.12. The large (A) and small (B) defects are planar point defect agglomerates (Bunkin and Nishnevitch, 1995; Becerro et al., 1999) in a (001) plane. A wide gap containing only small defects constitutes the mobility range of the wall at moderate fields (B). The crystal was glued to a supporting frame at one end. The AE setup was identical to the one described in Sect. 4.1. The filter bandwidth was 200 kHz to 2 MHz, a resonant type microphone (150 kHz) was used, the amplification was 54 dB, threshold 25 mV, for the AE and 43 dB, threshold 400 mV, for the BHP detection on

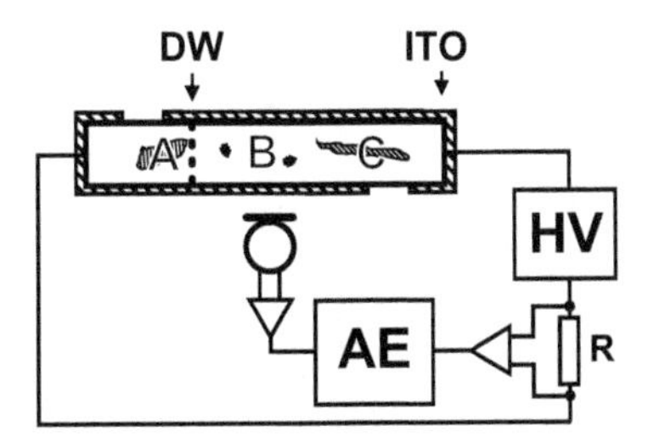

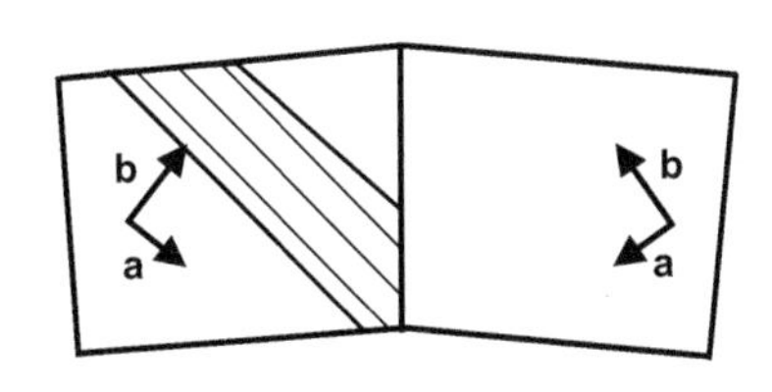

Fig. 4.13. Left: GMO crystal containing a single domain wall in its electrical circuitry. U-shaped In_2O_3:Sn-electrodes (ITO) are connected to a high voltage source (HV). The microphone beneath the sample, the serial resistance (R), the electronics (AE) and three regions of defects (A,B,C) are shown. The polarization directions are indicated. Right: Top view of the GMO crystal (Fig. 4.12) indicating the crystal axes orientations and resultant bending of the crystal (exaggerated)

a serial resistance (R=100 Ω, Fig. 4.13). A gap between the sample and the microphone was introduced to avoid the fracture of the sample. As displayed in Fig. 4.13(b), the orthorhombic a and b-axes are oriented with respect to the domain wall in such a way, that the bar develops a small angle at the location of the wall bending towards the shorter a axis (Newnham et al., 1969). The angle is sufficiently large to induce stresses that would fracture the crystal, if it were tightly bound to a flat microphone on the (110) plane. The gap between the sample and the microphone was acoustically bridged using liquid Fluorinert® (FC-77, 3M corporation), which as a liquid will only transmit longitudinal waves. Two crystal orientations were chosen, one with the crystallographic c-axis (AE || c-axis) and one with the [110] direction (AE ⊥ c-axis) normal to the microphone surface. Large wall jumps were visible in the optical microscope (25x) and recorded onto video tape (frequency 25 Hz). The wall was moved by bipolar triangular electric fields of rising maximum amplitude as shown in Fig. 4.14. Positive fields drive the wall across the range of small defects (B), negative fields towards the extended defect (A), high positive fields against another large defect (C, not seen in Fig. 4.12). (A) and (C) are only surmounted at very high fields not discussed here.

The first AE measurement (Fig. 4.14) was performed with the microphone cylindrical axis || to the crystal c-axis. Initially the wall is fixed by point charges due to aging (see Sect. 1.9). When breaking loose at V = -200 V, a large number of AE occurs ((1) in Fig. 4.14) showing spectra of type (c) in Fig. 4.16, but no coincident BHP. In the optical image no movement of the domain wall is seen yet. During subsequent cycles first BHP ((2) in Fig. 4.14) occur not coincident to an optically visible jump. For fields above +260 V (rate 26 V/s) numerous cascaded BHP occur along with optically visible large wall jumps. In this orientation AE of only very small energy are coincident with both the BHP and the optical jumps. A large portion of BHP does not

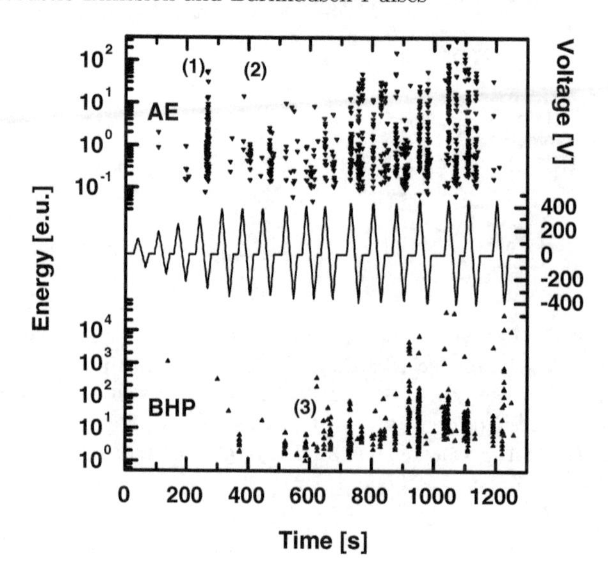

Fig. 4.14. Amplitudes of the acoustic emissions (AE) and Barkhausen pulses (BHP) plotted vs. time for a sequence of bipolar cycles. The corresponding electric field is shown. Crystal orientation is || c-axis

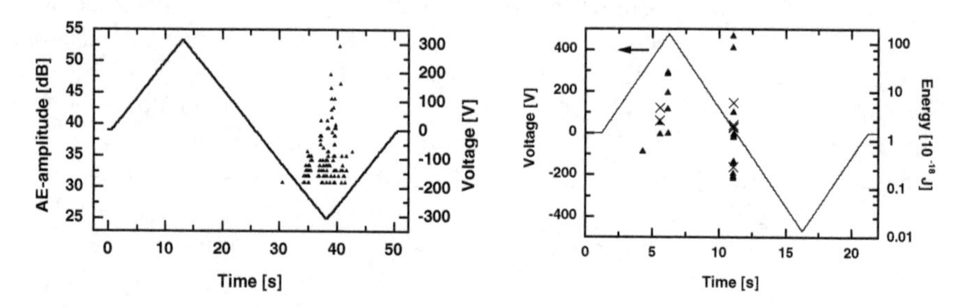

Fig. 4.15. Parts of the measurement shown in Fig. 4.14 at a higher time resolution. Left, first AE and, right, AE and BHP occurrence for optically observed simultaneous domain wall jumps

coincide with the optically visible jumps. All BHP are very brief, 1 to 5 μs. Long transients like seen in $BaTiO_3$ (100 ms, Chynoweth (1958)), constituting large polarization changes are only sometimes seen and follow the brief BHP. Additional non coincident AE occur having characteristic frequency spectra of type (e) (Fig. 4.16), when moving through the gap (B), or they are of type (d), when driven against defect (C).

The spectra (a)-(e) of the AE contain many frequency components. The resonant frequency of the microphone is beneath the filter cut-off frequency and thus all frequencies have to originate in the sample. Considering the size of the sources of AE in this crystal, the elementary jump process is probably not directly reflected by the electric signal leaving the microphone. The setup reacts as a resonating body, which is excited by the initial discontinuous event. The resonator can be understood in a similar sense as a church bell being excited by a bell-clapper. No matter what type of clapper is used and how hard it is hit against the bell, the bell will loudly ring only at its eigenfrequency. The peculiarities of the initial event can only be found in the high frequency components of the signal, for which the setup is not appropriate. In our case the sample is sufficiently decoupled from the microphone via the liquid Fluorinert not to form a joint resonator with the microphone. Thus the sample bar and parts of it constitute different resonators. Of the possible vibrations of the bar the lowest resonance is in bending. With the elastic constants of GMO (Höchli, 1972) the harmonics for vibrations along the c-axis generating pressure components towards the microphone for the bar clamped at one side are given by (Morse and Ingard, 1968): $\nu_1^{\|} = 1.7\,\text{kHz}$, $\nu_2^{\|} = 11\,\text{kHz}$, $\nu_3^{\|} = 31\,\text{kHz}$, $\nu_4^{\|} = 60\,\text{kHz}$, $\nu_5^{\|} = 100\,\text{kHz}$, $\nu_6^{\|} = 150\,\text{kHz}$, $\nu_7^{\|} = 208\,\text{kHz}$, $\nu_8^{\|} = 277\,\text{kHz}$, $\nu_9^{\|} = 355\,\text{kHz}$, $\nu_{10}^{\|} = 445\,\text{kHz}$, $\nu_{11}^{\|} = 543\,\text{kHz}$, $\nu_{12}^{\|} = 652\,\text{kHz}$, $\nu_{13}^{\|} = 770\,\text{kHz}$, $\nu_{14}^{\|} = 900\,\text{kHz}$. $\nu_7^{\|}$ through $\nu_{14}^{\|}$ fall into the detectable frequency range. The spectra do not only show these components, but there are others not originating from the bending motion. Particularly the events for traversing region (B) show different spectra each broad but seldom identical to Fig. 4.16(e). The additional frequencies are likely to be vibration modes of the domain wall as a membrane (Morse and Ingard, 1968). Unfortunately neither the mass per area density of the wall nor the tension on the wall are known quantities and a fit to the observed spectra will not yield a unique solution to determine both quantities and the integer pairs necessary to describe the harmonics. The thickness mode of the bar (2.8 MHz lowest harmonic) is beyond the detectable range of frequencies, but other possible modes are the longitudinal and torsional vibration of the bar (integer multiples of 53 kHz and 37 kHz respectively). Both are unlikely to be transmitted by the perpendicularly oriented microphone. Furthermore equidistant frequency lines should be observed in either case. The vibration of the wall is the most plausible source of the AE spectra. Thus, when stuck to the larger defects and particularly when travelling across the multitude of smaller defects, vibration motions of the wall are also excited.

In the second geometry the AE were recorded $\perp$ to the c-axis. Here the observed acoustic frequency spectra are less diverse. A clear single frequency (Fig. 4.16(b)) is seen for the AE coincident with BHP and the optically observable jumps. In this orientation large domain wall jumps will induce shear in the bar resulting in high amplitude pressure waves emitted perpendicularly into the contacting liquid. Of the calculated resonant frequencies: $\nu_1^{\perp} = 4.3\,\text{kHz}$, $\nu_2^{\perp} = 27\,\text{kHz}$, $\nu_3^{\perp} = 76\,\text{kHz}$, $\nu_4^{\perp} = 150\,\text{kHz}$, $\nu_5^{\perp} = 250\,\text{kHz}$, $\nu_6^{\perp} = 370\,\text{kHz}$, $\nu_7^{\perp} = 510\,\text{kHz}$, $\nu_8^{\perp} = 690\,\text{kHz}$, $\nu_9^{\perp} = 880\,\text{kHz}$, the frequency component $\nu_6^{\perp}$ is exactly reproduced in the AE spectra (Fig. 4.16(b)). The stable position of the domain wall is at 5.2 mm from the free oscillating tip of the

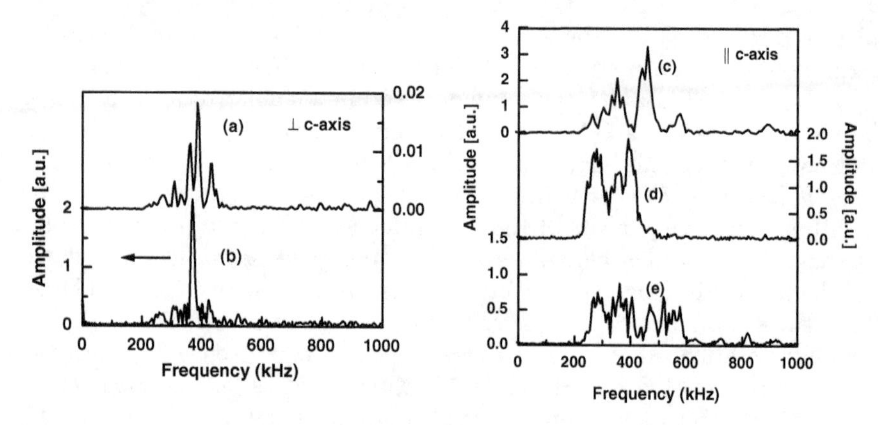

Fig. 4.16. Frequency spectrum of acoustic emissions ⊥ c-axis: (**a**) Coincident with a Barkhausen pulse and (**b**) non coincident; and ∥ c-axis: For the domain wall driven against a major obstacle (**c**), (**d**) and across the region with small obstacles only (**e**)

bar, the location of the anti-nodes of $\nu_3^{\perp}$ and $\nu_6^{\perp}$. Only $\nu_6^{\perp}$ falls into the detectable frequency range. The large domain wall jumps excite the bending vibration of the bar as expected. If not coincident with the large domain wall jumps, the AE in this geometry again show a different type of spectrum (Fig. 4.16(a)), which can not be explained by the bending vibrations only.

On a finer time scale it was often observed that some AE precede the BHP by 20-30 milliseconds. The wall seemingly loosens itself from the defect to an extent, sufficient to induce some vibrations before the entire liberation of the wall generates the high polarization changes. When truly coincident, the threshold crossings for AE and BHP differ by less than a few microseconds. AE preceding a subsequent cascade of coincident AE and BHP also show the spectra of type (a), where the term cascade denotes a sequence of briefly following events (30-200 microseconds between each).

Considering previous literature, AE and BHP were only observed when a domain wall is driven to an electrode edge (Zammit-Mangion and Saunders, 1984). At the edge high stresses and fields induce zigzag domain wall patterns appearing and disappearing under emission of BHP due to the locally induced high mechanical stresses and electric fields (Shur et al., 1986). This is a different mechanism than encountered here and will be discussed in the next section, where the nucleation and annihilation of entire domains becomes a dominant AE mechanism. The previously stated one to one time coincidence between AE and BHP in $Gd_2(MoO_4)_3$ (Zammit-Mangion and Saunders, 1984) can no longer be considered generally valid according to these results. Their result was probably due to unfortunate trigger conditions, which were entirely avoided in our set-up, as both channels trigger

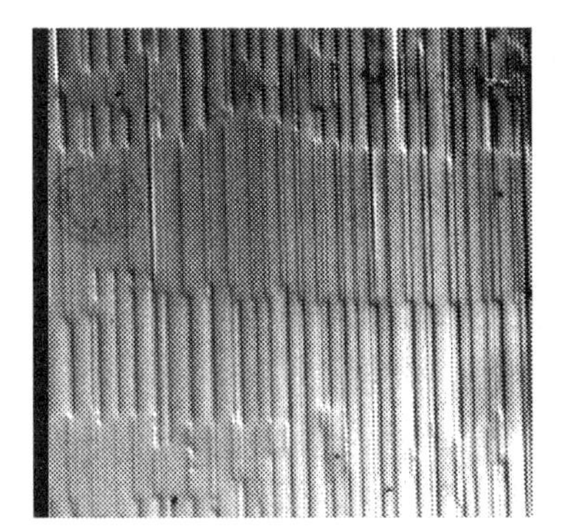
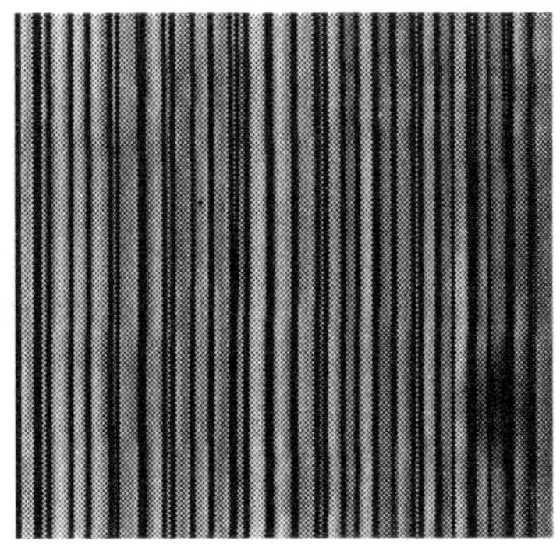

Fig. 4.17. Domain structure of the $BaTiO_3$ single crystal used for the acoustic emission tests, left before and right after the acoustic emission test. The electric field is applied left to right

independently and only later coincidence conditions are tested according to the common internal clock of the device.

For the single crystal GMO it can be summarized that bending vibrations of the entire bar and the excitation of different vibrational modes of the domain wall reacting as a membrane are observed. These events occur, when the wall is driven across or against defects in the crystal. The ultrasound spectra are characteristically different for coincident and non-coincident AE and BHP and for different defects involved. Furthermore, AE and BHP are not equivalent events in ferroelectric-ferroelastics as previously assumed, even in the case of a simple 180° domain wall.

4.3.2 The Perovskite Type Barium Titanate

Acoustic emission patterns from single crystalline $BaTiO_3$ were measured on commercial single crystals of $10 \times 10 \times 1\,mm^3$ size. The crystals contained domain patterns as shown in Fig. 4.17.

Essentially two important observations can be deduced from these data. For a sample that contains initially mixed striped *a*- and *c*-domains, which along the long edges of their predominant orientation exhibit further 90° domains, the application of an electric field along the sample width (10 mm side) will orient all these domains into the plane. This switching is initially clamped by some amount of mechanical and particularly electrical obstacles, which developed during sample aging. Initially every polarization change is accompanied by significant amounts of acoustic emission (Fig. 4.18). Fields well exceeding the later observed coercive field are needed to loosen the domains from their aged locations. After many bipolar cycles the text-book hysteresis develops. Now acoustic emissions only occur in the high field regions, when very little polarization changes are still encountered, Fig. 4.19. The large amounts of 180° switching around E_c are almost free of emissions.

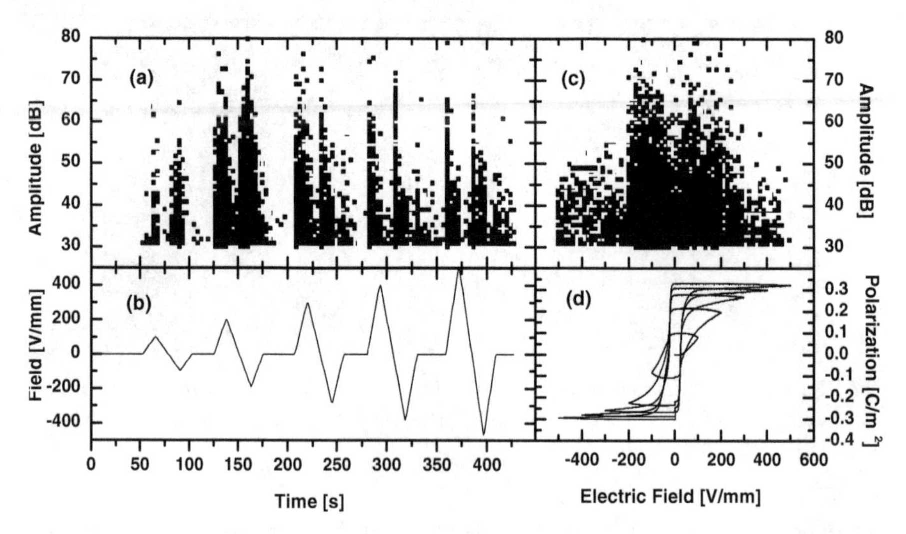

Fig. 4.18. The acoustic emission pattern for single crystal $BaTiO_3$ during five subsequent field cycles of rising bipolar amplitude. (**a**) Acoustic emission amplitude vs. time, (**b**) electric field vs. time, (**c**) acoustic emission amplitude vs. electric field, and (**d**) polarization hysteresis

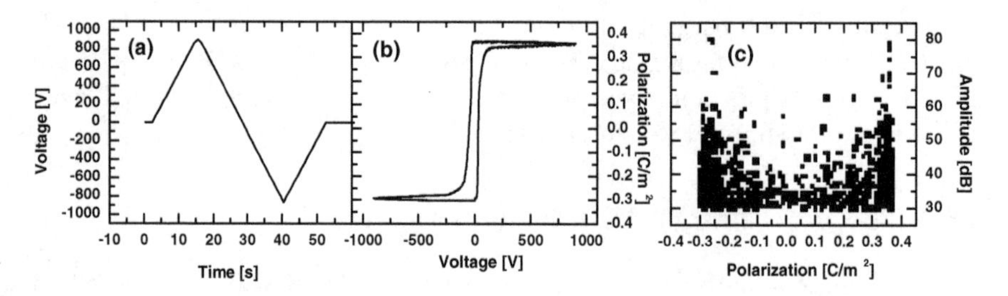

Fig. 4.19. The acoustic emission pattern for single crystal $BaTiO_3$ at high fields and polarization. (**a**) Time dependent voltage (10 mm sample length) (**b**) polarization hysteresis, and (**c**) acoustic emission amplitude vs. polarization

Acoustic events predominantly occur each time larger fields are needed to generate further polarization changes, like when the steep switching in the polarization hysteresis becomes somewhat slower for rising fields above a polarization value of about $0.15\,C/m^2$. Thus all microscopic events that generate acoustic emissions are due to driven domain switching events that necessitate a certain amount of extra energy provided by the external driving forces, here the electric field.

The crucial message of this measurement for the fatigue studies and measurements in ceramics is that it is definitely the domain switching which will generate the dominant contribution to the acoustic emissions at high

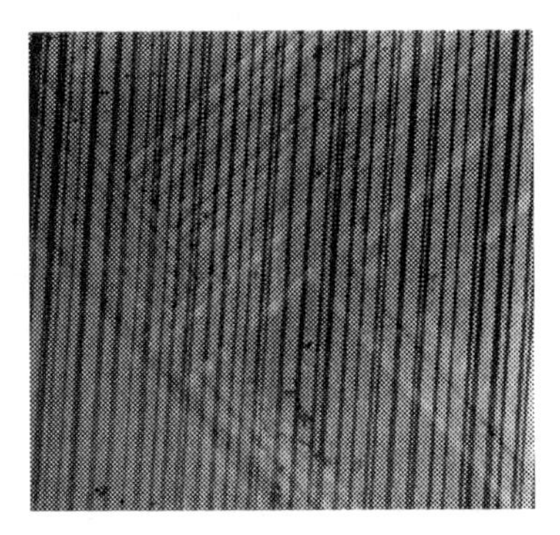

Fig. 4.20. Domain structure of the $BaTiO_3$ single crystal before the acoustic emission test containing mechanically induced needle domains

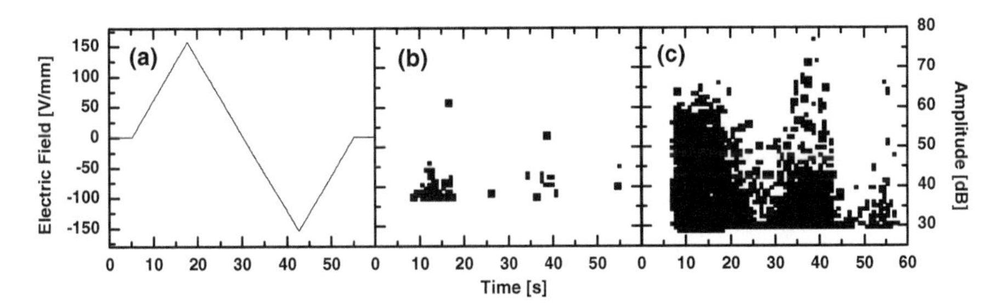

Fig. 4.21. Acoustic emissions and Barkhausen pulses from needle domains in $BaTiO_3$, which had been induced by mechanical stresses. (**a**) Applied fields, (**b**) acoustic emissions at this field for a sample that had been de-aged like in Fig. 4.18, and (**c**) acoustic emissions after introduction of needle domains

fields in an unfatigued ferroelectric ceramic. Microcracking is an effect of secondary importance concerning acoustic emissions in ferroelectrics, which was also suggested from the energy estimates in ceramic PZT (2% La-doped) in Sect. 4.2.2. There, the microcracking only constituted the low energy events.

A second set of measurements on the identical sample was performed after large amounts of needle domains had been introduced mechanically. The needle domains penetrated the sample at 45° to the outer sample faces. Again an external electric field was applied. The annihilation of the needle domains generated enormous amounts of high amplitude acoustic emissions (Fig. 4.21, (c)), much higher than the switching data obtained before at this voltage (Fig. 4.21, (b)). Thus the annihilation of domain walls is a process liberating very large amounts of elastic energy. In the later stage of fatigue domain wall creation and annihilation may become a significant part of the domain switching process inducing the increased amounts of high amplitude acoustic emissions observed after fatigue. This may occur besides the already discussed increased influence of obstacles on the mere domain wall motion.

5 Models and Mechanisms

The present chapter is essentially the discussion of the multiple factors influencing fatigue in ceramic ferroelectrics. It is thus in large parts also a review of relevant literature. The intention is to provide more than just a mere presentation of a single model, but the attempt to correlate the vast body of knowledge on particular aspects of fatigue to the relevant mechanisms.

A model will be presented on the aspect of point defect agglomeration. Except for one or two experimental results in the literature all experimental observations including some of the so far only vaguely explained data can be understood in the framework of the microscopic agglomeration mechanism. The macroscopic observations then are a direct consequence of agglomeration.

5.1 Structure of Fatigue Models for a Polycrystalline Ferroelectric

The theoretical description of fatigue in polycrystalline ferroelectrics has so many facets, that a single model description will inherently be insufficient. Thus, different scales of effects have to be addressed and modelled. The structure of such models has to encompass scales from the individual point defect to the macroscopic boundary conditions of the sample. Already at this point it is clear that such a task is beyond the scope and feasibility of this work. Nevertheless, the different relevant parameters will be addressed as completely as possible and some suggestions for an improved modelling will be given for particular aspects.

(1) The smallest scale relevant to fatigue modelling is the individual point defect. This point defect does not match the electronic and elastic structure of the host lattice and will therefore alter the local ordering within the unit cell. The first effect is in most cases the loss of ferroelectricity of this unit cell.
If the defect is charged, meaning that the corresponding electrons and holes have moved into the conduction or valence band, respectively, it will exert a considerable force onto its neighborhood, often termed relaxation of the neighborhood in physical chemistry. This force in turn will reorient the polarization of the neighboring unit cells towards the charge.

Fortunately, the polarizability of ferroelectrics is extremely high, so the forces will decay fairly rapidly into the volume of a ferroelectric domain. Nevertheless, a finite volume will be so strongly oriented towards the defect, that it can no longer or only partially participate in the ferroelectric switching process. The local polarization around the defect may thus differ considerably from the polarization of the surrounding domain. As will be shown, electronic as well as ionic point defects each contribute to the fatigue phenomena.

(2) The next scale is constituted by an agglomerate of defects. Such an agglomerate may be as small as a single dipole or larger units of agglomerated ions as was shown in Chap. 3. For an isolated non charged dipole the forces will decay even more than in the case of an isolated charged ion, as is well known from electrostatics (Jackson, 1975). Nevertheless, the influence of the dipoles may be large, because their influence is mediated through the interaction with the polarization of an entire domain. Thus domains may be clamped, even by distributed dipoles.

(3) The growth of such agglomerates will primarily happen on the length scale of the domain. For this to occur, diffusion has to take place. Three aspects are relevant in ferroelectrics, the anisotropy of the diffusion constant within the domain, which will generate point defect gradients even without external fields present. Then the external field will induce an additional drift force on the ionic species. The third aspect is the influence of polarization on the diffusion process itself. The positions for the defect within the unit cell are not equiprobable. Therefore the net drift of ions along or against the polarization direction will be different. This has been shown to be a large effect. Agglomerates, which have grown to some larger size, are likely to strongly interfere with the domain system. This effect is large, as will be shown in this chapter.

(4) The domain system itself is a response of the ferroelectric to the development of the depolarizing field, when the spontaneous polarization is formed. It reduces the total free energy of the solid. This energy is electric or elastic in nature. Thus defects, if they are mobile, will take part in this energetic stabilization process and move in order to reduce the overall energy of the system. Aging due to diffusion across larger sections of the domain occurs.

(5) The role of grain boundaries in polycrystalline materials is evident. They are responsible for a reduced polarizability, reduced domain widths in smaller grains and local mechanical stresses in the grain after the material has been cooled beneath its Curie point. These local fields and stresses participate in the fields acting on the ionic defects. Under certain doping conditions, grain boundaries may even develop a proper charge state, which in some cases will be of only one polarity. In turn a very high barrier to electronic and ionic motion across the grain boundary develops for a particular sign of charge.

(6) So far the considerations have been static. Upon field reversal the domain structure is usually completely altered. During the domain wall motion, the wall or the domain interior of inverse polarization may interact with all the above mentioned defects. The motion will be altered on different time and size scales according to the different defects encountered. On the other hand, the moving domain walls may carry along a finite number of point defects during their motion. Temporarily, the order of the structure within a grain may be heavily distorted. Comparably many point defects may be moved due to the induced drift with respect to diffusion. This will be termed convection in the following, but this terminology is not reflecting a general use of it.

(7) Most commercial PZT compositions are close to the morphotropic phase boundary. The strong interaction of the domain system with defects may stabilize certain defect configurations, which are more favorably placed in either the rhombohedral or tetragonal structure. Thus, crystallographic phase changes may be induced.

(8) As a polycrystalline material is considered, the external field will only be acting according to its projection onto the crystallographic axes of the particular grain. Certain smaller defects may impede domain wall motion in grains with the smallest projection of the external field onto any of its permitted polarization axes. Defects of similar size might still be considered energetically small in such grains where one crystallographic axis is favorably oriented parallel to the external field.

(9) The external electrodes are metallic and it is known that metal - semiconductor interfaces will generate Schottky barriers. The resultant space charge layer may alter the net field experienced in the bulk, if enough mobile electronic charge carriers are available.

(10) The ferroelectric electrode interface furthermore becomes a barrier to the diffusion of ionic defects or a source of new defects in the ceramic. In the first case a space charge layer will be formed by the ionic defects. These are also more likely to become located at the electrodes, because their diffusion through the metal is much reduced.

Not all these effects are similarly encountered in thin films and fatigue modelling will therefore differ to some extent between thin films, thick films and bulk ferroelectrics.

In the following sections the relevance of the above mentioned constituents of a complete fatigue model will be evaluated with respect to existing and new experimental data. It will be shown, that fatigue occurs at several different degrees of severity. The most fatal mechanism is microcracking. A similarly bad mechanism is the formation of hard agglomerates, which will turn out to be oxygen vacancy clusters that agglomerate to planar defect planes. Many experimental observations match this mechanism. Several effects of fatigue are not as stable and can be recovered by high fields or thermal treatment. These are generally space charge effects. Two types of space charges have to be separated, ionic defects and electronic defects. Furthermore their locations have to be specified, because they lead to either local or global effects. The ionic defects are less mobile and constitute a fairly stable state of fatigue, even those which do not agglomerate. Electronic space charges are the most mobile species and can be excited by light or higher temperatures. They are responsible for easily recoverable fatigue effects. As also some deep trapping levels have been observed, electronic defects may play a larger role in fatigue than initially anticipated.

5.2 Band Structure

The primary information necessary for the description of fatigue is the knowledge of the charge state of the defects involved. Different from the high temperature equilibria investigated in the defect chemical studies discussed in Sects. 1.8 and 5.3 the charge state at room temperature is not as evident.

Unfortunately, Fridkin (1980)'s book on semiconductor properties of ferroelectrics was written so early (Russian original in 1975) that electronic properties of PZT compounds were not yet fully investigated. Even since then, the data on the PZT band structure are not as abundant, as one might have expected from its technical relevance.

Kala (1991) gave a collection of known energy levels of defects in the band gap of PZT. According to models and experimental studies of disordered solids electron as well as hole states exist just beneath the conduction or above the valence band, respectively. Kala assigns these states to fluctuating bond lengths like in the case of disordered solids. As for PZT this is probably only true in the vicinity of grain boundaries, where some local disorder will occur. In the bulk such states are more likely to be polaron states. Kala gives a long list of possible defects in PZT. A detailed set of energies associated with doped PZT was given for Mn- and Fe-doped PZT. For the series of La-dopant compositions $Pb(Zr_{0.6}Ti_{0.4})O_3 + 0.015\ MnO_2 + y/2\ La_2O_3$ the energies of the defects are given in Figure 5.1. In the case of Fe-doping ($Pb(Zr_{0.6}Ti_{0.4})O_3 + x/2\ Fe_2O_3 + y/2\ La_2O_3$, $x = 0 - 0.5$ and $y = 0 - 0.1$) he determined transitions with energies: 0.9 eV, 1.2-1.3 eV, 1.6 eV, 1.65-1.7 eV, 2.4-2.5 eV, 2.75-2.8 eV, 2.9-3.2 eV, 3.4-3.8 eV, but did not assign the individual transitions to particular defects. The effect of such localized defects on the ferroelectric properties was then discussed in the same work (see Sect. 5.5).

Robertson et al. (1993) and Robertson and Warren (1995) calculated the band structure for the compound system PZT using the tight-binding approximation. As the lower and upper band edges are determined by the Pb 6*s*- and O 2*p*-electrons and 6p-electrons respectively, little change of the band gap is observed across the composition range from $PbTiO_3$ to $PbZrO_3$ (3.45 eV to 3.65 eV). At the morphotropic phase boundary a value of 3.6 eV is calculated. Compared to these results from the lead based perovskites, the difference in band gap between the completely filled electron shells of barium

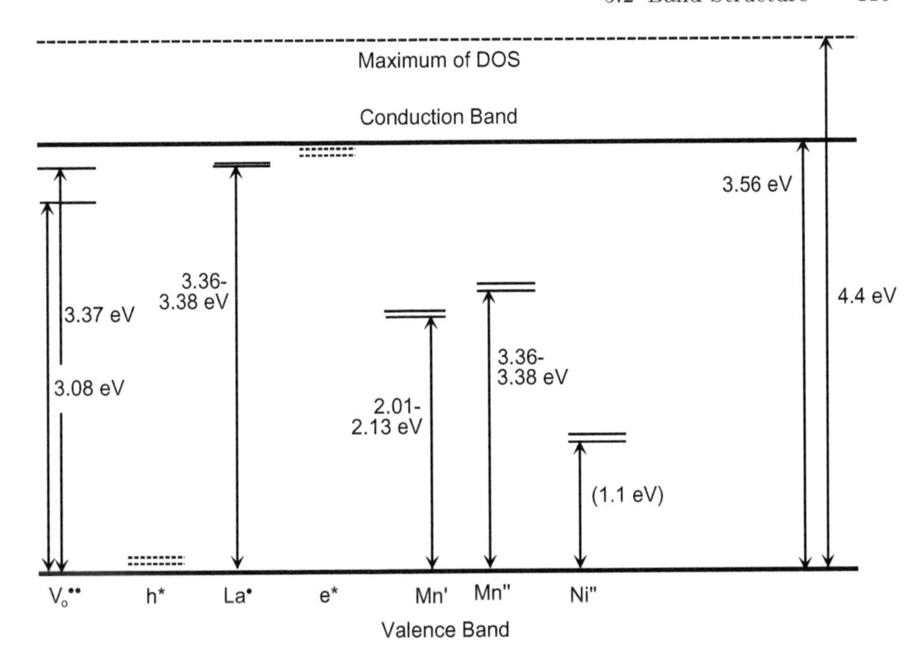

Fig. 5.1. Band structure of $Pb(Zr_{0.6}Ti_{0.4})O_3 - 0.015MnO_2 + y/2\,La_2O_3$ for different concentrations $y = 0$ to 0.045 of La: $V_O^{\bullet\bullet}$, $h^{\bullet}$, $La_{Pb}^{\bullet}$, e', Mn'_{Ti}, Mn''_{Ti} (Kala, 1991). The value for Ni''_{Ti} is from $SrTiO_3$ (Waser and Smyth, 1996). (DOS = density of states). h^* and e^* denote polaron states in order to differentiate them from the generally valid notations $h^{\bullet}$ and e'

in $BaTiO_3$ and $BaZrO_3$ is huge, because the band edges are now determined by the Ti and Zr 4d states. The band gap changes from 3.4 eV for $BaTiO_3$ to 5.0 eV for $BaZrO_3$ (Prokopalo, 1979).

Miura and Tanaka (1996a) calculated the effect of La-doping on the band structure of PZT forming PLZT. The major influence arises due to the $4f$ electrons of La. These levels are located slightly beneath the Ti $3d$ states forming the conduction band edge and thus slightly decrease the band edge energy, if their concentration is high enough.

In an early study Miller and Glower (1972) determined the conduction of electronic charge carriers using Hall-effect measurements and temperature dependent conductivity measurements. They arrived at a defect for electrons associated with the oxygen vacancy at 0.22 to 0.28 eV beneath the conduction band edge, which is a slightly lower range in energies than given by Kala (1991) in Fig. 5.1. Miller and Glower (1972) further determined the electron mobility to be $\mu = 0.2 \pm 0.12\,\mathrm{cm^2/(V\,s)}$.

5.3 Point Defects and Dipoles

5.3.1 Concentration

The first consideration is the starting concentration of defects in the material. The first and obvious concentration of defects are the dopants. Thus we here talk about 2% lanthanum on the one hand and 8% doping by the ionic pair antimony and nickel in ratio 2:1 on the other.

The fourth relevant defect is the oxygen vacancy . Its concentration is determined during the sintering of the material at elevated temperature and will cease to vary significantly upon cooling at about 800 K (≈ 500°C) (Waser and Smyth, 1996). It is formed according to the general equilibrium equation:

$$O_o \rightleftharpoons \frac{1}{2}O_2 + V_o^{\bullet\bullet} + 2e^- . \tag{5.1}$$

The formation of vacancies is then determined by the Gibb's free energy of formation

$$\frac{[e^-]^2 \, [V_o^{\bullet\bullet}] \, p_{O_2}^{1/2}}{[O_o]} = \exp\left(-\frac{\Delta G}{kT}\right) \tag{5.2}$$

for one vacancy.

As was pointed out in the introduction (see p. 21), donor doping will slightly reduce the effect of fatigue cycling (Chen et al., 1994a). As the donor doping reduces the number of $V_O^{\bullet\bullet}$ and possibly increases the number of cationic defects, $V_O^{\bullet\bullet}$ is definitely less wanted in the microstructure than cationic defects.

In a very insightful experiment Brazier et al. (1999) showed that the lowest total conductivity yields the best fatigue resistance. They varied the concentration of free charge carriers in thin film PZT (55/45 and 75/25) by varying the oxygen partial pressure $p_{O_2} = 10^{-3}$ to 1 bar across the conductivity minimum. Their experiments demonstrated two facts, first the critical role of the oxygen stoichiometry and secondly that fatigue increases irrespective of a higher electron or hole concentration. At moderate differences of external partial pressure from the value at the conductivity minimum, that the number of ionic species is generally considered to stay constant and only their charge state to vary (Kingery et al., 1976). For the rhombohedral composition the best fatigue resistance occurred at higher values of p_{O_2} than in the tetragonal composition. The value was 200 mbar being the ambient atmospheric partial pressure.

A surprising effect concerning the concentration of mobile charge carriers in the case of acceptor doping was found for Ni-doping in $SrTiO_3$ (Waser and Smyth, 1996). For Ni-concentration between 0.05 and 0.8 at% the conductivity of $SrTiO_3$ did not change in the quenched state. A conductivity change according to a slope of about +1 in a log-log plot would have been expected for the defect reactions (1.17) and (5.3):

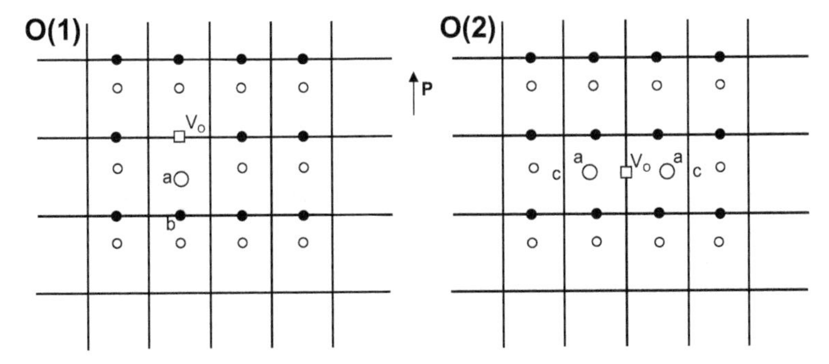

Fig. 5.2. The distortion of the polarization in the perovskite type lattice due to a single charged substitutional point defect. Of all lattice ions only the chains of titanium ions (open circles) and oxygen ions (full circles) are shown. The distortions are different for vacancy positions at O(1) sites and at O(2) sites (see also Fig. 5.3). The ions that do not match the pattern of the surrounding domain are marked by the large circles. A charged domain wall develops at location b, partially charged domain walls at locations c

$$\mathrm{Ni_{Ti}} \rightleftharpoons \mathrm{Ni''_{Ti}} + 2\mathrm{h}^{\bullet} \ . \tag{5.3}$$

Waser and Smyth (1996) explained this inconsistency by allowing for a high degree of association of both types of vacancies, which would turn the $\mathrm{V_O^{\bullet\bullet}}$ immobile:

$$\mathrm{Ni''_{Ti}} + \mathrm{V_O^{\bullet\bullet}} \rightleftharpoons \{\mathrm{V_O} - \mathrm{Ni_{Ti}}\}^{\times} \ , \tag{5.4}$$

where $^{\times}$ denotes a net charge-free state in the lattice. Such dipoles may be induced by certain cycling conditions, which in a way collect freely distributed point charges and drive them to associated entities as $\{\mathrm{V_O} - \mathrm{Ni_{Ti}}\}^{\times}$.

5.3.2 Defects and Microdomains

Localized charges interact with their immediate environment by reorienting the neighboring unit cells. As pointed out by Brennan (1993) and later modelled in the framework of first-principles pseudopotential total energy calculations by Park and Chadi (1998) (also see the discussion of the interaction with domains, p. 141) the immediate neighborhood of a defect ion like $\mathrm{V_O^{\bullet\bullet}}$ deforms according to two possible types of constellations depicted in Fig. 5.2 . The first one concerns a vacancy position in Ti-O-Ti chains along the polarization direction of the surrounding domain (O(1)), the second vacancy positions perpendicular to this chain (O(2)). In these calculations it became clear, that the influence of the vacancy is directed in space and not the entire environment will relax in a larger fashion towards the defect.

The influence of an isolated charge in a somewhat more remote location is given by Coulomb's law $F \propto 1/(\epsilon_o \epsilon_r r^2)$, which à priori is a fairly far reaching influence. As ferroelectrics are highly polarizable even within a single crystalline domain and without any domain wall contributions, the forces will drop off much more than in low-ϵ materials.

Dipole fields drop off even more rapidly $F \propto 1/(\epsilon_o \epsilon_r r^3)$ and they will most likely not reorient their immediate environment (Jackson, 1975; Robels, 1993). Nevertheless, a strong interaction with the surrounding domain will be present.

In the above sense microdomains were found in the perovskite $KTaO_3$ (Grenier et al., 1992) induced by oxygen vacancies. They were detected in luminescence spectra originating from doubly charged Ta$''$ denoting a Ta^{3+} ion in the vicinity of microdomains. The oxygen vacancies are responsible for the formation of the microdomain as well as the charge state of the tantalum ion.

A study of different PLZT compositions irradiated with neutrons showed that the irradiation will induce a broad phase transition like in relaxors. Relaxors are known to have a nanodomain structure at the origin of the relaxor effect (Cross, 1994) and a similar effect seems to be present here. The hysteresis in the irradiated sample very much resembles those of fatigued samples. Thermal annealing similar to the results in Sect. 2.3 will return the sample to its unfatigued state.

Similar results, but on a different scale are those concerning space charge layers at electrodes. As was already found in the late 50'ies (Fatuzzo, 1962), space charge accumulates underneath metal electrodes in $BaTiO_3$. The region underneath the electrodes was shown to contain a considerable amount of differently oriented domains no longer switchable by the external field (Williams, 1965). This effect was much less pronounced with electrolyte electrodes.

It is clear from the literature data presented in this section, that the formation of localized domains around single or agglomerated ionic defects, particularly oxygen vacancies is highly probable also in PZT.

5.3.3 Localized Electron States

In the highly polarizable ferroelectric materials a fairly strong interaction of an electron with its elastic environment may be anticipated. Thus, small and large polaron[1] states have been identified in PZT. An effective mass of

[1] A small polaron is a highly localized electron state inducing a strong deformation of its neighboring environment. It generates its own energy minimum by deforming its elastic environment. For any polaron state to exist, the time of interaction with phonons has to be long enough so that at least one full wavelength of the phonon can interact with the electron. Thus, only slow electrons (low k-value) can form polarons. Along with the elastic deformation around a polaron an induced polarization of the environment occurs. A large polaron does not interact as strongly with its environment and is therefore more mobile. In

$m^* = 3m_e$ was determined for PZT, which is larger than e.g. in KCl, which is known representative of small polarons (Hellwege, 1988).

Early on in the investigations of conduction in PZT the majority carriers were assigned by Prisedsky et al. (1978) to be small polaron states. In their study the partial pressures of PbO, p_{PbO}, and O_2, p_{O_2}, were externally imposed on the PZT samples up to 1000°C. For low p_{PbO} the samples showed a strong off-stoichiometry, while for high p_{PbO}, almost stoichiometric PZT is formed. The measurement of the Seebeck-coefficient yielded predominant p-type conductivity, which was also matching the p_{O_2}-dependence of conductivity (exponent +0.25) at high temperatures. No dispersion of a.c. conductivity was found above 400°C, which was assigned to the fact that at high temperatures the grain boundaries have little influence on the total conductivity. This is actually very different from the data by Vollmann and Waser (1994) on $SrTiO_3$, where a significant step in conductivity was observed between 50 kHz and 1 MHz even at 600°C indicating that two distinct parts of the microstructure determine the conductivity (see p. 136). In PZT the activation energy for conductivity decreased continuously from 1.0 eV at room temperature to 0.6 eV at 950°C indicating a change in the predominant defect equilibrium. The activation energy for hole conductivity is 0.45 eV for the temperature range 600°C to 1000°C. The activation energy for the hole mobility drops from 0.26 eV at 600°C to 0.15 at 950°C. By considering the estimated overlap-integrals of the electron states, the phonon-spectra and the mobility data, they deduced that the conduction mechanism in PZT is small polaron conductivity. Considering the electroneutrality condition

$$2[\mathrm{V}_{\mathrm{Pb}}''] + n = 2[\mathrm{V}_{\mathrm{O}}^{\bullet\bullet}] + p \tag{5.5}$$

they arrive at a dominating defect concentration of $[\mathrm{V}_{\mathrm{O}}^{\bullet\bullet}] \approx [\mathrm{V}_{\mathrm{Pb}}'']$ in the p-conductivity region. Using previous data on the maximum concentration of $\mathrm{V}_{\mathrm{Pb}}''$ (2.5%) and their own thermopower data they estimated that a maximum of around 6% $\mathrm{V}_{\mathrm{Pb}}''$ can be compensated by holes, the remainder has to be compensated by oxygen vacancies. They did not consider other donor dopants which are part of the minerals they used for sample preparation and my participate in the charge equilibrium.

From analogy with tungsten, molybdenum or tellurium oxides Güttler et al. (1995) used Raman spectra to estimate the effective mass m^* of a polaron in PZT. They also arrive at $m^* = 3m_e$ which is much heavier than in many other oxides. Thus a higher polarizability of PZT also around a single electron seems to exist, which can actually be expected from the high macroscopic polarizability of the ferroelectric perovskites.

Not all electron states are found to be localized polaron states. This was also discussed with reference to a vast body of literature in the same paper

semiconductors with wide band gap also holes can be considered hole polaron states of inverse charge, because they interact with their elastic environment in a similar manner (Hellwege, 1988; Tsuda et al., 2000).

by Güttler et al. (1995). The clustering of oxygen vacancies actually leads to Anderson localization of the electronic states, which is a particular form of polaron (Fridkin, 1980; Hellwege, 1988). Thus around clusters of oxygen vacancies a high number of localized polaron states exist. Only a fraction of all electrons generated according to equation (5.1) are liberated into the conduction band. The authors mention fractions of one (0.5) out of the possible two electrons for oxygen vacancy concentrations below (above) 8%. The clustering of oxygen vacancies in these works is confined to the surface of the sample, which is distinctly different from the data presented in Chap. 3.

The discussion of polaron and other localized electronic states significantly alters the degree at which the defect structure for acceptor doped samples is unbalanced. Usually, e.g. for iron doping, two iron ions are needed to equilibrate one oxygen vacancy. If by the polaron mechanism, one electron state is coupled to the oxygen vacancy generating it, the total charge balance is maintained and only one vacancy is generated for each iron ion. When the oxygen vacancy is doubly charged ($V_O^{\bullet\bullet}$), the liberated electron may be kept in the closer vicinity of the $V_O^{\bullet\bullet}$ without actually recharging the vacancy as a nearby localized polaron and the effective dipole moment of the defect dipole can be compensated in the closer environment. The screening can also occur somewhat more remotely by the polaron when acting as a screening charge for larger numbers of oriented Fe^{3+}-$V_O^{\bullet\bullet}$ dipoles. Such larger distances can be mediated through the polarization in the ferroelectric domains.

In PIC 151 nickel is a primarily doubly charged acceptor ion and the matching of one oxygen vacancy with one nickel is perfect. If the $V_O^{\bullet\bullet}$ on the other hand do maintain a polaron localized in their vicinity, the material as a whole will become softer in character than the balanced doping with $NiSb_2O_6$ (PIC 151) suggests from a purely ionic defect equilibrium point of view. The soft behavior is actually found for PIC 151 and an explanation using the polaron argument seems reasonable.

Robertson et al. (1993) and Warren et al. (1994b, 1996d) have discussed the localization of holes in PZT by investigating Pb^{3+} and Ti^{3+} centers. Their study concerned PZT sintered in oxygen rich atmospheres, where the excess oxygen leads to cation vacancies and p-type conduction, which is the usual case. The question of localization of the holes is relevant with respect to the fatigue mechanism, which will become evident in the following sections. The predominant hole conduction at high temperatures (see Sect. 1.6.2) is frozen in at room temperature due to deep traps localizing the conduction holes. When illuminated, these holes are brought into the valence band from where they recombine with the lead ions to form the active EPR-center Pb^{3+}. After excitation, the number of such centers decreased only within the time range of hours (room temperature). The particular properties of these Pb^{3+} centers was assigned to the fact, that the valence band edge in the band structure is constituted by 45% Pb 6s states and 55% O 2p states. This allows the formation of metastable polaron states slightly above the valence band edge localized at the lead ion. Similarly, electrons can become localized at Ti-ions creating Ti^{3+} centers, which actually exhibit an even longer half-life in the range of many hours to days.

5.3.4 Defect Dipoles

As mentioned in the introduction the alignment of point defect dipoles can explain the stabilization of certain domain orientations (Robels et al., 1995b). The dipoles are mostly constituted by an acceptor ion and an oxygen vacancy $V_O^{\bullet\bullet}$ (A'-$V_O^{\bullet\bullet}$ or A"-$V_O^{\bullet\bullet}$) or possibly by a double vacancy pair V_{Pb}''-$V_O^{\bullet\bullet}$. Robels et al. (1995b) and other works by the Arlt group (see p. 23) explained aging and hard doping in the framework of orienting these defect dipoles along the local polarization of the domain. With the knowledge of this model and the fact that domain stabilization by hard doping meets all the macroscopic experimental evidence, it was surprising to find that the dipole alignment of ionic defect pairs was *not* found locally by EPR measurements (Warren et al., 1995a, 1996b). Only external electric fields and additional UV illumination or additional elevated temperatures just beneath the Curie point were sufficient to effectively orient the dipoles and generate anisotropy in EPR spectra.

5.4 Ion and Electron Motion

5.4.1 Diffusion Within a Domain

Non-equilibrium thermodynamics describes the material behavior, when thermal equilibrium has not been reached. Thermal equilibrium in this circumstance determines a state, for which all fluxes of extensive microscopic variables vanish on a mesoscopic and macroscopic scale, i.e. equilibrium is reached in temperature, stress, and chemical potential homogeneous throughout the microstructure. Before such equilibrium is reached, fluxes of extensive variables traverse the solid, which are entropy, chemical species and in a certain sense strains, which entails stress relaxation. Driving forces are the gradients of the intensive variables, temperature, stress, and chemical potential and the external forces. Electric field and stress are the two external forces considered here, because no magnetic fields are present and gravity is neglected.

A brief but very well written introduction to the transport processes relevant in ionic solids is given in Kingery et al. (1976). For the electrical part a similarly well written introduction was summarized by Hench and West (1990). A detailed theoretical description of atomic transport is found in Allnatt and Lidiard (1993), which is an excellent treatise of this subject. It uses much of the classical book on non-equilibrium thermodynamics by Groot and Mazur (1962), but also incorporates the particular properties of defects in solids and their treatment in the framework of non-equilibrium thermodynamics. The different fatigue models delineated in the introduction use the general results from Hench and West (1990), particularly the treatment by Yoo and Desu (1992c), on which most subsequent models are based.

All treatments so far, except for a few works from Russia (Kanaev and Malinovskii, 1982; Sturman, 1982; Khalilov et al., 1986), have not been concerned with the fact that the material develops the spontaneous polarization. It has to be emphasized here, that the term *polar* crystal, as used in most of the textbooks on this subject, denotes the fact that ionicity in the material is high and it is used to distinguish it from the metal case. This is distinctly different from a *polarized* crystal, which we will be concerned with here. The most general case of a polarized crystal is a pyroelectric, for which a spontaneous polarization develops upon cooling. In the following the general scheme for the development of a diffusion model will be followed according to the above mentioned books, but the particular aspects of pyroelectrics will be included. Even though a ferroelectric is generally not macroscopically pyroelectric due to the development of the domain structure, each individual domain represents a pyroelectric crystal.

In usual considerations of the transport processes the generalized fluxes of different extensive variables are not independent. The well known Onsager relations show that particle and entropy fluxes are interrelated. Particle fluxes will be driven by gradients in temperature, gradients in chemical potential and gradients in stress. Furthermore, in the case of piezoelectrics, the strains and stresses are related to the polarization and electric field via the piezoelectric coefficients. These interrelations have to be kept in mind in the forthcoming paragraphs.

5.4.2 Anisotropy and Directionality of Diffusion

The following section represents a discussion of the theoretical implications of the fact that diffusion in a pyroelectric is inherently directional, which has so far not been incorporated into a three dimensional form of a diffusion model. A difficulty that arises in discussing the flux of point defects through a pyroelectric is to arrive at a relation between a microscopic defect model and a macroscopic average law for the ion fluxes that includes the geometry of the problem. In polycrystalline ferroelectrics the orientation of the electric field is arbitrary with respect to the polarization and the crystal axes of a domain and a complete geometrical description is needed.

Geometrical Considerations

In an ionic crystal the flux $\mathbf{J}^{(l)}$ of a charged defect of type (l) is essentially driven by the electric field and the gradient of the chemical potential. In Ohm's law, the current density is proportional to the applied electric field. Thus, in a simple first approximation, the current density is given by

$$J_i^{(l)} = \sigma_{ij}^{(l)} \left(E_j - \nabla_j \mu^{(l)} \right) . \tag{5.6}$$

$\sigma_{ij}^{(l)}$ is the second order conductivity tensor for ionic species (l) and ∇_j the j-th component of the gradient of the fractional chemical potential $\mu^{(l)}$ (for the precise definition see Allnatt and Lidiard (1993)). If $\sigma_{ij}^{(l)}$ is not equivalent to a scalar, this form already incorporates the full anisotropy of the defect current. In the case of a ferroelastic, which does not develop an internal polarization, i.e. which maintains an inversion center upon cooling through the phase transition temperature, this law is fully sufficient. As the local polarizations in an antiferroelectric reside in such proximity to each other, the influence of the single unit cell polarization will average out over the neighborhood of a defect ion. Thus equation (5.6) is also valid in the case of an antiferroelectric. The easiest coordinate system to write down coefficients $\sigma_{ij}^{(l)}$ of $\bar{\bar{\sigma}}^{(l)}$ is the proper coordinate system of the crystal structure, namely that coordinate system yielding highest symmetry with respect to the crystal axes. In the case of a tetragonal crystal this becomes:

$$\sigma_{ij} = \begin{bmatrix} \sigma_{11} & & \\ & \sigma_{11} & \\ & & \sigma_{33} \end{bmatrix} \tag{5.7}$$

implying that the tetragonal distortion occurs along the 3-direction. The σ_{ij} will have an exponential temperature dependence according to some activation energy (Gibb's free energy (Allnatt and Lidiard, 1993)), a point of later discussion. It is important to note that (5.6) will already induce a gradient in defect concentration between 90° domain walls and other regions in the bulk and is thus a potential mechanism for aging.

The problem that now has to be worked into the geometrical structure of this law is the fact that the defect current strongly depends upon the polarity of the external field with respect to the spontaneous polarization. This was shown by measurements in $LiNbO_3$, where differences by a factor of 10 in current densities were found along and opposite to the polarization direction (Kanaev and Malinovskii, 1982). Sturman (1982) developed a uniaxial model for defect diffusion in a pyroelectric following the usual scheme for microscopic models which will be outlined in the subsequent paragraphs of this section. Khalilov et al. (1986) on the other hand used a similar difference of a factor 5 in conductivity found in a poled 7/63/35 PLZT ceramic to support an electronic model of conductivity assuming the electronic defects to be the most relevant mobile species in the system. They modelled the material to be a series of p-n-junctions, each double layer of this p-n-junction constituted by a single unit cell. For now it is only relevant, that this again was a uniaxial model and that a strong difference in conductivity was anticipated along and against the polarization direction in PLZT. We thus cannot ignore this fact in modelling fatigue, if we are discussing PZT and related compounds. This effect will be disclosed to be particularly relevant in the vicinity of already existing agglomerates (see Sect. 5.8.5). The effect is equally relevant in aging.

The problem now is to reconcile this with the conductivity tensor $\bar{\bar{\sigma}}^{(l)}$. The major problem is to generate a form that includes polarization from a structure that in itself is not polarized. When $\bar{\bar{\sigma}}^{(l)}$ is used, there is no difference in numbers if $\mathbf{E}$ changes sign, $\mathbf{J}$ will also just change sign, but not its numerical value. Thus, $\bar{\bar{\sigma}}^{(l)}$ is not sufficient to describe our problem.

Two approaches are possible now, first a model assigning an assumed generalized force to the material. It acts just like an external electric field. The easiest form of this is to assume that the material is not yet in charge equilibrium with its environment and the depolarizing field, which is primarily constituting the pyroelectric effect, also drives the ionic motion in the material.

A second approach may use the material property polarization itself. Sturman (1982) used the polarization (actually the local dipole moment) times the electric field to describe the difference in the jump rate of ions into neighboring potential wells in his uniaxial model. The dipole moment itself was used as a measure for the sign dependence and amplitude of the ionic current. This model yielded plausible numbers and is thus worthwhile to be considered for a generalization. An easy incorporation of this contribution is a scalar product of E and P. It takes care of the fact that the directionality difference only occurs along the direction of polarization and also includes the direct proportionality to P. If P becomes zero, the directionality should vanish, but a finite conductivity, possibly anisotropic, should pertain. The following form includes all these requirements:

$$J_i^{(l)} = \left[\sigma_{ij}^{\circ} + a_i p_j\right] X_j^{(l)} \qquad \text{or}$$

$$\mathbf{J}^{(l)} = \left[\bar{\bar{\sigma}}^{\circ} + \mathbf{a} \otimes \mathbf{p}\right] \cdot \mathbf{X}^{(l)} \qquad \text{with} \tag{5.8}$$

$$\mathbf{X}^{(l)} = \mathbf{E} - \nabla \mu^{(l)} \tag{5.9}$$

where $\mathbf{p}$ denotes the spontaneous polarization, $\mathbf{a}$ the directionality coefficient, and $\mathbf{X}^{(l)}$ the generalized force. Due to the geometrical structure, $\mathbf{a}$ is a vector quantity and $\mathbf{a} \otimes \mathbf{p}$ a second order tensor. The form of equation (5.8) ensures that the conductivity remains finite for a vanishing spontaneous polarization and that the general orientation of $\mathbf{p}$ does not have to coincide with the coordinate axes of the unit cell coordinate system. The projection of the electric field onto the polarization yields the sign and amplitude necessary for describing the differences in current density in the two opposing directions and the dyadic product yields the necessary structure of a conductivity tensor. $\mathbf{a}$ may itself be a function of $\mathbf{p}$. As most quantities in ferroelectric theory can be deduced from some Landau potential, $\mathbf{a}$ is likely to be a polynomial in $|p|$.

Experimentally it is not very easy to disentangle $\bar{\bar{\sigma}}$ and $\mathbf{a} \otimes \mathbf{p}$. The easiest way would be to find two compositions close to each other, one being antiferroelectric and the other ferroelectric, but both for identical defect concentrations. The antiferroelectric compositions should not contain the second term in (5.8) and the difference in measurement should be significant. On the other hand the effects determined by Sturman (1982) and Khalilov et al. (1986) are so large, that actually the second term will be dominating in real pyroelectric systems.

In the PZT compositions considered in this work, the description (5.8) is sufficient for the material description. For sake of completeness, the most general case of an additional anisotropy induced due to the polarization is considered. The crystal structure induces an anisotropic ionic conductivity $\bar{\bar{\sigma}}$. Upon formation of the spontaneous polarization, this polarization may induce an additional anisotropy correlated only to its particular orientation in space. The additional anisotropic conductivity will couple to the electric field to yield the current density but has to maintain the directionality of equation (5.8).

$$J_i = \left[\sigma_{ij} + A_{ik}\widetilde{E}_k p_j\right] E_j \quad \text{or}$$
$$\mathbf{J} = \left[\bar{\bar{\sigma}} + (\mathbf{A} \otimes \mathbf{p}) \cdot \widetilde{\mathbf{E}}\right] \cdot \mathbf{E} \; . \tag{5.10}$$

The unit vector $\widetilde{\mathbf{E}}$ along the field direction is chosen to maintain a linear material law. Again $\mathbf{A}$ may be a function of $|p|$, which is likely a polynomial as discussed above.

This section is relevant with respect to aging as well as fatigue. First we briefly consider aging. The net flux of a particular ionic species is directed along the direction of the spontaneous polarization. When this flux of ions reaches a planar boundary like a domain wall or a grain boundary, the net current will decrease, particularly at grain boundaries (Waser and Smyth, 1996). Ions and thus net true charges accumulate at the grain boundaries. This accumulation process continues until the net ionic current due to the directionality of the domain balances the electric potential gradient newly created by the accumulating ions. This process only stops, once the total chemical potential $\mu^t = \mu^{(l)} - ez^{(l)}\phi_{\text{defect}}$ has reached equilibrium. $z^{(l)}$ in this case is the charge number of the defect species (l), e the elementary charge and ϕ_{defect} the potential created by all defect charges already accumulated at the domain boundary (wall or grain boundary). Once fully complete the newly accumulated ions generate an additional field, which adds to the external field and represents at least a local offset field. Such an approach will be further discussed in Sect. 5.8.4.

Attempt Frequency

For PZT σ°_{ij} in (5.8) describes an anisotropy induced due to the increasing tetragonality beneath the Curie point. It could thus equally well be used in the context of an antiferroelectric. The simple anisotropy of the diffusion represented by σ°_{ij} without considering the polarization can be related to the attempt frequency. This frequency is provided by the phonons in the crystal. Thus, a straight forward linear approximation of σ°_{ij} in the coordinate system provided by the crystal yields

$$\sigma^{\circ}_{11} = \sigma^{\circ}_{22} \propto a \cdot \sqrt{\frac{c_{\alpha\beta}}{\rho}} \tag{5.11}$$

$$\sigma^{\circ}_{33} \propto c \cdot \sqrt{\frac{c_{\alpha\beta}}{\rho}} \tag{5.12}$$

with $c_{\alpha\beta}$ the elastic constants, ρ the density of the material and a, c the dimensions of the tetragonal unit cell. The attempt frequency describes the fact that an ion moves in its potential well at a certain oscillation frequency. The energy for surpassing the potential barrier is the thermal energy and included in the exponential term, while the attempt frequency denotes the rate at which an ion attempts to surmount a particular potential barrier. The attempt frequency in a solid is given by the frequency of the thermal phonons. These in turn are determined by the elastic constants of the material. Most ionic compounds like NiO or MgO, for which many data are known, but also $SrTiO_3$ are simple cubic structures and the attempt frequency is isotropic according to the elastic constants. It is emphasized here, that the macroscopic elastic constants are not appropriate in the case of ferroelectrics and the local anisotropy is much higher than the macroscopic one. While the anisotropy in the case of the oxygen atoms will be determined by the rigidity of the crystal as a whole and be fairly low and similar to that of the macroscopic crystal, the anisotropy for the cations and particularly the titanium ion is much higher. This will be similarly true for isovalent defect ions. For the titanium ion the anisotropy will be about as high as the anisotropy of the dielectric constant extrapolated to the phonon frequencies, if one assumes that the major charge transfer constituting the local dipole moment is due to the displacement of the titanium ion. The attempt frequency tensor becomes

$$\bar{\bar{\nu}} = [\nu_{ij}] = \begin{bmatrix} \nu_1 & & \\ & \nu_1 & \\ & & \nu_3 \end{bmatrix} \tag{5.13}$$

for the tetragonal composition and thus has the same symmetry as (5.7). If a coordinate system along the cubic [111] direction is chosen, the attempt frequency tensor takes on the same form as (5.13) for the rhombohedral composition. Transforming it to the cubic coordinate system yields a fully filled symmetric tensor. Using the anisotropy of the dielectric constant for the tetragonal lead zirconate titanate near the morphotropic boundary: $\epsilon_3^S = 399$, and $\epsilon_1^S = 612$ (Jaffe et al., 1971) one obtains a ratio of the attempt frequencies of 1.5. Raman spectra from PZT (95/5) (almost pure $PbZrO_3$) show a pronounced soft phonon peak, which shifts from $80\,\mathrm{cm}^{-1}$ at 10 K to $54\,\mathrm{cm}^{-1}$ at room temperature Hafid et al. (1992). This is equivalent to an attempt frequency of $1.6 \cdot 10^{12}$ Hz, slightly lower than the commonly assumed 10^{13} Hz for oxide materials (Kingery et al., 1976).

Another important point is the significant difference in mobility of certain defect dipoles with respect to isolated defects. Vacancies move at a much higher rate than the lattice ions, because they do not need a vacant neighboring site to jump to. Any lattice site of the same crystallographic sub-lattice is accessible to the vacancy. Thus, if a dopant ion generates a neighboring vacancy on the same sublattice, the mobility of the ion will be significantly enhanced. This can happen with donor ions on the lead site in PZT. Considering the equilibrium reaction (1.12) on p. 8, two lanthanum ions are in electrochemical equilibrium with one lead vacancy. As agglomerates of two lanthanum ions associated with one lead vacancy can not occur for every vacancy due to the reduced configurational entropy, a significant portion of the lanthanum ions will not be associated with a lead vacancy. It is usually assumed that defects form pairs and not triplets, but triplets are not excluded here. So there will be three different mobilities associated with these defect associates, an essentially immobile isolated lanthanum ion, which will diffuse at the rate of the intrinsic lead ions, a defect pair, which will be effectively as mobile as an isolated lead vacancy (Kingery et al., 1976), because it always has its vacancy associated with it, and the triplet, which will again be fairly immobile.

For acceptors the story is different. The electrochemical reaction (1.14) yields the highly probable defect pair Ni''_{Ti} - $V_O^{\bullet\bullet}$, but the vacancy is located on a different sub-lattice. The Ni-ion can not jump to the oxygen site, and thus the mobility of the defect pair is only an orientational one. These orientation effects have been discussed thoroughly by many authors (see Sects. 1.9 and 5.3.4).

Similarly, antimony will be as mobile as the intrinsic titanium ions, because the associated lead vacancy also occupies a separate sub-lattice (Reaction equation (1.13)).

Another effect of the incorporation of acceptors is the reduction of the tetragonality. The anisotropy of the average unit cell dimensions was determined by X-ray diffraction to drop down from $(c/a-1) = 0.001$ to 0.0004 for an Fe^{3+}-concentration of 2% (Hagemann, 1980). This effect may be closely related to the effects seen in our measurements due to fatigue. The reduction of the average unit cell size may be correlated to the effectively stronger influence of acceptors in the bulk of the PZT grains. The reduction by $5 \cdot 10^{-3}$ in our samples (see Sect. 3.5) may thus just be the effect of the increased acceptor character of a certain number of ions in the material. But equally well more complicating pictures may be involved considering the interaction of domains and domain walls with the growing defect agglomerates.

Ion Mobility

Waser (1991) determined the oxygen mobility from conductivity data in hot pressed ceramic $SrTiO_3$ using impedance measurements in the time and frequency domain. This technique permitted the separation of contributions from the grain boundaries and the bulk grains. He obtained values for the bulk mobility of $\mu_{V_O^{\bullet\bullet}}(513\text{K}) = 3$ to $7 \cdot 10^{-9}\,\text{cm}^2/(\text{V}\cdot\text{s})$ without consideration of any association. According to:

$$\mu_{V_O^{\bullet\bullet}} \propto \frac{1}{T} \exp\left(\frac{-E_A}{kT}\right) \tag{5.14}$$

a thermal activation energy between 1.005 eV and 1.093 eV from low temperature data compared to 1.1 eV for high temperature data (Chan et al., 1981) was obtained. Waser (1991) could furthermore show, that ionic diffusion dominates the charge transport in the quenched state at room temperature, because it was virtually independent of the oxygen partial pressure ($p_{O_2} = 10^{-11}$ to 10^5 Pa) during equilibration. The effect of the quenching rate (5 K/s to 30 K/s) was smaller than the average experimental scatter in the data.

Activation Energy and its Anisotropy

In $BaTiO_3$ a thermal activation energy of 0.70 eV was found for the fatigue process (Kudzin et al., 1975). This may be closely related to the energy needed for the diffusion of the charged defects responsible in $BaTiO_3$ for inducing the fatigue effect.

Another effect, which was shown to be crucial concerning the motion of ions in polarized domains is the anisotropy of this jump barrier. This was explicitly shown in the case of cage motion of the oxygen vacancy around a Ni-acceptor ion (Arlt and Neumann, 1988). It is reasonable to assume that similar anisotropies in the energy barrier exist for the motion of an isolated oxygen vacancy, not adjacent to an acceptor ion trapping it. This anisotropy may even exceed the aforementioned anisotropy of the attempt frequency in the preceding paragraph. Such an assumption is also supported by the calculations of the microscopic environment of the oxygen vacancy according to Park and Chadi (1998) and Fig. 5.2. Unfortunately no numbers are available for this effect.

5.4.3 Drift and Convection

Another aspect, which has so far not been considered in literature is the influence of moving domain walls on point defects. Particularly, needle domains carry a large electric field at their tips, which is effectively a charged domain wall. This high charge may momentarily induce an enhanced ionic current, while the domain passes the defect ion. The extremely high local fields are certainly able to liberate a vacancy from its present position. If only one cell is traversed in one bipolar switching cycle, 10^8 cycles may carry an ion on average some 10^4 unit cells away, which is once across an average grain. This effect may thus be large at certain locations in the grain, where needle domains are generated and travel certain distances across the grain. At the endpoints of such travel paths a certain type of agglomeration process may take place, which is actually different from the one discussed later on in

Sect. 5.8, but which may nevertheless constitute one of type of space charges, which are responsible for parts of the fatigue effects observed.

5.4.4 Directionality of Electron Motion

According to the model by Khalilov et al. (1986) not only the ionic but also the electronic conduction is highly directed. The model assumes a series of p-n junctions formed in highly doped and compensated ferroelectrics (Sandomirskii et al., 1982) and uses PLZT as an example. From a polynomial expression of the polarization they arrive a difference in p-n junction width along and against the externally applied electric field. This in turn induces an effective electron (or hole) current directed along the polarization direction. A rectification factor 5 was found for poled PLZT at an effective electric field of 200 V/mm, which is well beneath the coercive field. Under light illumination the rectification actually decreased during their experiments.

A similar effect was observed by Warren and Dimos (1994) (see a further discussion of their results in Sect. 5.5, p. 132). They investigated the pinning of domains in $BaTiO_3$ under d.c. fields. Aside from fatigue, the most effective reduction of the amplitude in the polarization hysteresis was achieved by applying a field inverse to the prior poling direction just beneath the coercive field and by illuminating with UV light of the band gap energy. In this case an almost total reduction of switchable polarization was observed. Two things are crucial about this, firstly the effect is maximized, when the field is just opposite to the polarization and, secondly, when the band gap energy is chosen. Thus, electron-hole pairs are generated by the UV light, they are excited into highly immobile states, because they can fix the polarization, and the electrons and holes drift apart under the local field. The strongest driving force for the charge separation effect is given, when the fields due to the spontaneous polarization and the external field add up. This kind of charge separation and pinning is different from the agglomeration encountered in our experiments. The charge carriers separated by the combined UV - bias - treatment are not hard agglomerates, because they can be recovered by applying high electric fields along with UV illumination or by cycling. This type of pinning is moreover similar to the offset found in our experiments. Both are induced by the effectively unidirectional motion of ions or localized electrons (polarons). The pinning occurs essentially at grain boundaries, or in the case of unipolar loading also at domain boundaries (Al-Shareef et al., 1997; Warren et al., 1997c), because these are fixed during the d.c. treatment.

5.4.5 Average Conductivities

Leakage current measured after prolonged times is essentially the only current, which can not be confused with slow relaxation due to domain reorientation. In thin films (sol-gel, RTA sintered) Cho and Jeon (1999) showed that at room temperature relaxational behavior is observed. Whether this is truly relaxational current from free mobile defects or the relaxational reorientation of domains is not clear. For higher temperatures on the other hand strong differences occur depending on the atmosphere during sintering. For N_2-sintered samples conductivity hardly changes with time, while in fully oxidized films the conductivity rises fairly abruptly at a certain moment. The authors explain their finding by an increased electron or hole density by space charge redistribution at the electrodes due to the electric field. A more likely understanding of the abrupt increase in leakage current are the mechanisms discussed in detail by Waser et al. (1990a,b); Baiatu et al. (1990); Waser (1991); Vollmann and Waser (1994); Vollmann et al. (1997); Waser and Hagenbeck (2000). After long times of d.c. loading, large amounts of charge carriers accumulate at the grain boundaries, until the fields across the grain boundaries become so huge, that charge carrier injection into the next grain sets in.

5.5 Interactions Between Point Defects and Domain Walls, Screening, Space Charges, and Domain Freezing

General Effects

Different vocabulary has been used in the literature to essentially describe the same effect. It concerns the interaction of electrically charged defects with the domains. Screening, internal fields, offset fields and similar terms have been used. Each of these has been used to either denote global or local effects, with the transition between both scales being vague, particularly in thin films, where everything happens essentially on the same scale.

To my opinion actually a different approach should be chosen, particularly in view of the observed offset-polarization, also termed imprint polarization in thin films. The different perspective is that primarily, two different types of charge carriers have to be considered, namely ionic and electronic charges.

Secondly a differentiation should be undertaken between those charge pile-ups, that will only modify the external electric field, but still permit domains to move and switching to take place, and those entities, which after pile-up will no longer permit the domain system in their vicinity to be reoriented (domain freezing, domain clamping). As will be seen in the discussion of different fatigue models, the amount of the latter species is determined by the number of hard agglomerates for example. Shur et al. (2001b) in their models have just introduced a volume fraction of non-switchable polarization and used this as a fitting parameter to the actual experimental data without reference to a local mechanism (see the further discussion in Sect. 5.8).

One important result of such a differentiation is the experimental difference between offset-fields and offset-polarizations. Offset-fields still permit the domain system to move fairly freely. They usually only shift the polarization hysteresis along the field axis. These offset fields are generated by charges not directly interacting with the domain system, but only contributing to the overall fields experienced in the microstructure. Preferred locations for such pile-ups are grain boundaries or the external electrodes. At these locations the domain wall motion is either already hindered for certain domains like at grain boundaries or the external field is directly modified at or near its external origin. The strong difference between the single crystal remnant polarization and the polycrystalline remnant polarization shows that a large part of the domains in ceramics is not switched, anyhow. In the case of $BaTiO_3$ this means more than 50% loss of polarization. The mobility of domains within each grain is not significantly modified due to these charges only modifying the external fields. The 50% loss encountered anyway with respect to the single crystal is not modified by additional grain boundary charges.

The second aspect is the offset-polarization. It means that some domain walls have ceased to move. If only a reduction in switchable polarization is observed, this does not necessarily mean that an offset has developed. If the orientation of the fixed domains is random on average, the effect will be a mere reduction in polarization and strain hysteresis. Once the orientation is no longer random, a macroscopic offset-polarization is observed, like in our experiments (Sects. 2.3, 2.6, and 2.7). Under certain loading conditions, like in the case of unipolar cycling in Sect. 2.6, a combined effect of an offset-field and an offset-polarization was observed.

A third differentiation may be introduced, when the motion of domain walls is sufficiently clamped at certain external fields but not at higher ones (different degrees of clamping). This essentially yields a distribution of coercive fields. The hard agglomerates are responsible for this effect. They are entities truly interfering with the domain wall movement, representing finite, but considerable energy barriers to their motion.

Literature

Robels et al. (1995b) gave a summary of the possible deformations of ferroelectric hysteresis loops due to inactive ferroelectric layers. High dielectric layers underneath the electrodes will render the hysteresis loop more slanted, internal bias fields are needed to shift the hysteresis loop along the electric field axis and the reduced electric field in the ferroelectric layer lets the measured coercive field appear larger. A second type of layer not in series, but parallel to the ferroelectric phase leads to only a slight slant of the hysteresis curve. They extended this simple layer model to paraelectric inclusions or inclusions, where the polarization is fixed, and found that inclusions can be replaced by a set of thin layers parallel and perpendicular to the electrodes and yield the same effects. This arrangement actually does not reduce the maximum polarization reached, if only the volume of the ferroelectric part is considered. Thus a reduction of polarization will not be induced by inclusions or dielectric layers except via the absolute volume fraction of these. The drastic polarization reduction during fatigue thus necessitates another origin, while effects generating slanted hysteresis curves can easily be explained by their model assuming non-ferroelectric dielectric layers or inclusions.

Another point of view was taken by Pan et al. (1992c). Under an electric field the Gibbs free energy for a double well potential becomes asymmetric leading to the polarization switching. It is induced due to the thermally stimulated switching according to the probability

$$p = \nu \exp\left(\frac{-Q}{k_B T}\right) \tag{5.15}$$

with $Q = Q_o - E \cdot P_s$. In a fatigued material one potential well is permanently lowered by point defects or space charges. The electric field is not sufficient to raise this lowered potential to sufficiently raise the total energy for switching this domain. The polarization freezes. Pinning under bipolar driving is a self stabilizing effect, because small fractions of the polarization are kept from switching due to local defects during the second cycle.

Dimos et al. (1994) have given the most comprehensive discussion of the influence of reordering of electronic defects under UV-light and bias conditions.

Illumination of a transparent PLZT 7/65/35 ceramic at a field of about 70% of $+E_c$ with band gap light results in a reduced amount of switchable polarization (60% height of the original P-E loop) and a polarization-offset. The switchable polarization is reduced to 25% of the initial polarization after 5 minutes exposing time, while the coercive voltage remains identical. The reduction of switchable polarization and the offset are stable up to bipolar voltages of $\pm 400\,\mathrm{V}$ (250 μm).

The reduction of switchable polarization very much depends on the absolute value of the bias field applied at illumination. Up to voltages of 2.5 V (E_c= 4 V, PZT 50/50, 800 nm, *thin film*, Pt-electrodes) the switchable polarization drops to 60% of its initial value. The effect of biasing is strongest, when electric fields opposing the present polarity are applied. In fully saturated materials, the additional illumination does not reduce the amount of switchable polarization.

Voltage biases at saturating field levels do not reduce the switchable polarization, but induce a significant voltage offset (PLZT-film 6/20/80, 1 μm thick, $V_s = \pm 15\,V$, UV 365 nm). Optical writing at $\pm P_r$ induces a voltage offset in bulk PLZT 7/65/35.

A film of PLZT 2/60/80 (1.06 μm thick) exhibiting a significant offset voltage after a UV illumination at the negative saturation voltage (- 15 V) can be re-centered by polishing off the first 150-200 nm on the film top. For the reverse polarity applied during UV illumination, no effect of polishing is anticipated.

The following statements can be derived. As all the changes in hysteresis are generated by UV light, it necessarily concerns electronic charge carriers. These become sufficiently located within the film or the grain structure, because they remain very stable during subsequent hysteresis cycles. Once liberated from their primary location, the electronic charge carriers migrate considerable distances through the material. In the ceramic samples 85% of the polarization can be suppressed, but only a layer thickness of 10 μm is illuminated by the incident light due to absorption (sample thickness 250 μm). These charge carriers then become trapped at locations, where they match the local depolarizing fields much deeper in the sample. As polishing off a consecutive number of layers from the ceramic surface does not alter the induced offsets, the resulting charge distribution in the microstructure occurs throughout the ceramic. The strongest imprint effects occur just beneath the coercive field. This is consistent with our observation that the local fields driving ion and electronic motion are highest in the case of the external field opposing the local depolarizing field of the domains (see Sect. 5.8.2 and 5.8.5).

In the case of the samples illuminated at saturation, the domain system is highly oriented. Thus all charges will become located at similar locations and their global effect will add up. This effect is particularly pronounced in columnar thin films (film thickness = grain size). All charges accumulate underneath the electrodes. The effect is an offset-field, because the localized charge carriers now modify the external field without significantly suppressing the domain wall mobility. In columnar thin films the reduction of switchable polarization arises, when space charge develops in the grain boundaries, which are essentially perpendicular to the electrode faces. Now the polarization is fixed perpendicular to the field direction. For only partial switching a considerable number of charged domain walls exists, which trap the free charge carriers immediately. The dominant trapped charge carriers in the samples were holes and the dominant driving force for charge migration is the depolarizing field, even if only residual fractions exist in the film or ceramic. The fact that polishing off the top layer after UV writing at saturation yielded different results depending on polarity shows that the offset-field is constituted by localized electrons. These must occupy sufficiently deep traps to be stable even after high field treatment. The major reason, why the freezing of domains in the bulk near E_c is so much more effective, lies in the fact that for the local depolarizing fields oriented perpendicular to external electrodes no free charge carriers are provided at the outer premises of the sample to compensate the local field. Thus many more free carriers are attracted to charged domain walls (or grain boundaries) in the bulk of a ceramic.

As a result of the electronic structure of point defects in the band gap of PZT, Kala (1991) also discussed the influence of localized charge carriers on the properties of ferroelectrics. A particular result of interest with respect to our data of the thermal liberation of frozen domains in Sect. 2.3.2 were the measurements on the thermally stimulated short-circuit currents. In poled and aged samples without any dopants the maximum current expectedly occurred at the Curie point T_c (500 °C), but a considerable current also existed at 360 °C of opposite polarity. In the case of 1% Fe-doping (Fe_2O_3) the latter maximum was found at 300°C, and only minor currents at T_c. For 1% Mn-doping (MnO_2) current maxima of equal amplitude existed in both polarities with a seemingly reduced Curie point. Pure donor doping with 2% La yielded currents only at 300 °C and co-doping with equal amounts of Mn^{4+} and La^{3+} yielded the same single current maximum event at lower temperatures (250 °C).

Another study anticipating the general relevance of free electronic charge carriers in the compensation of ionic defects was conducted by Peterson et al. (1995). They fatigued thin PZT films (53/47, 0.2 µm thick, Pt-electrodes) to 10^9 cycles and then measured the hysteresis loops with and without illumination. The illuminated films show considerably higher saturating and remnant polarization. Films sintered at 725°C showed some relaxation, until the polarization re-grew from 65% to 80% of the initial polarization, while the film sintered at 875°C also exhibiting higher maximum polarization values showed hardly any delay after light illumination.

Waser (1991) estimated a space charge layer thickness in the $SrTiO_3$ and evaporated Au or Ni/Cr-electrodes to be about 300 nm thick and that of grain boundaries to be 100nm (Neumann and Arlt, 1986a).

5.6 Electrodes

In the following sections it is assumed that the general quality of the electrodes is good, i.e. the surfaces were clean before application of the electrodes and suitable firing techniques are used to avoid effects of local electrode tips yielding singularities of field and immediately entail the failure due to strain and field mismatch between neighboring grains (Pan et al., 1989; Jiang et al., 1992).

A general discussion of the effects of electrodes on material properties of ferroelectrics can be found in the books by Fridkin (1980) and Scott (2000). Here only a brief summary is given.

Schottky Barrier and Semiconductor Diode Effects

Like in other semiconductors ferroelectric perovskites are subject to the development of Schottky diodes at metallic electrodes (Fridkin, 1980). A description of Schottky contacts to ferroelectric perovskites was given by Yoo et al. (1993). Their behavior is essentially the same as with other semiconductors, when the local fields due to the ferroelectric are included in the consideration.

As the electrodes in our samples do not dominate the bulk behavior, their importance in the fatigue mechanism is only considered secondary and a further discussion of the detailed electronic structure of the Schottky contact to a ferroelectric will not be given here. Only the effect of the silver diffusion into the sample is of relevance as was already discussed in Sect. 3.2.

Diffusion Barrier

As was briefly mentioned in Sect. 1.10, the diffusion of ionic species across metal electrodes is highly limited. Particularly oxygen is basically not soluble at room temperature. Oxygen vacancies will thus accumulate beneath the electrodes and form a space charge layer. This effect is particularly important in thin films (Scott et al., 1991), because the thickness of such layers is in the same order of magnitude as the sample thickness. During 10^9 bipolar fatigue cycles on 150 nm thick PZT films, a significant reduction of oxygen content in PZT was observed by Auger spectroscopy beneath a Pt electrode, while the effect was not as pronounced underneath gold electrodes (Scott et al., 1991). Another effect observed by Scott et al. (1991) was that the oxygen content was reduced already at some distance from the electrodes. This yields two regions of PZT, that have different semiconductor properties, one, the common PZT in the bulk is of p-type, while the zone near the electrodes becomes a n-type conductor. In a certain sense the sequential diode model by Khalilov et al. (1986) which is used to explain the unidirectional current may even have its microscopic origin at the grain boundaries where similar pile-up effects may occur, if these pile-up effects are partially blocking the ionic motion in the sense of Sect. 5.7.

Similarly the fatigue effect was previously directly correlated to the blocking of the oxygen vacancy diffusion across the metallic electrode by Peterson et al. (1995) (see also the results by Kundu and Lee (2000) on p. 29 and Dawber et al. (2001), Sect. 6.4.4). They interpreted this in the identical way, with the PZT being a n-type semiconductor due to the large number of available electrons from the oxygen vacancies according to equation (1.17). All commonly encountered PZT compositions are p-type conductors with low carrier concentrations, unless strongly donor doped, due to the natural abundance of lower valence ions in minerals. If the number of oxygen vacancies increases considerably in certain regions of the sample, this region will turn from p- to n-type conduction. Peterson et al. (1995) argued that the Fermi-level in PZT shifts towards the PZT conduction band edge and the work needed to excite an electron from the Fermi-level in the Pt-metal into the conduction band of PZT decreases considerably. The authors observed that white light was sufficient to modify the hysteresis response of PZT after fatigue, but had little influence before fatigue. The white light provides sufficient energy to excite electrons from the highest energy levels in Pt into the conduction band of PZT, but does not contain sufficient UV light contributions to actually bridge the intrinsic band gap of PZT. The enhanced number of free carriers re-charges the oxygen vacancies (equation 5.16), which then loose at least half of their binding charge to fix the neighboring domains.

$$V_O^{\bullet\bullet} + e' \rightleftharpoons V_O^{\bullet} \tag{5.16}$$

As the recovery of the hysteresis loop is only partial, the authors assume that fatigue is constituted by more than one mechanism. Thermal excitation will re-ionize the $V_O^{\bullet}$ to $V_O^{\bullet\bullet}$, with the electrons most likely moving back to the Pt-metal, and thus re-establishing the fatigue state after some aging time, because the binding energy for the electron at the $V_O^{\bullet}$ is only around 0.2 eV.

5.7 Grain Boundaries

Another important barrier to ion diffusion in ferroelectric ceramics is the grain boundary in the perovskite type titanates. Neumann and Arlt (1986a,b) were the first to recognize the crucial relevance of the grain boundaries also in undoped titanate perovskites. Maxwell-Wagner relaxation was observed, which means that the ceramic behaves like a sequence of resistors and capacitors in series, the capacitor being the grain boundary and the resistor the grain bulk. Under the application of electric field the ions and possibly partially bound electronic charges are moved towards the grain boundaries, where their motion is stopped due to the capacitor character of the grain boundary. A charge pile-up occurs, which in turn generates a fairly high field across the grain boundary. An activation energy of the mobile species of 0.7 eV was assigned. The data by Neumann and Arlt (1986a) were further extended by Waser (1991), Vollmann and Waser (1994), Hagenbeck et al. (1996), Vollmann et al. (1997) and Vollmann and Waser (1997). Electrical conductivity measurements in the frequency and time domain showed the grain boundaries to become invincible barriers to the diffusion of positively charged point defects, $V_O^{\bullet\bullet}$ and $h^{\bullet}$, in acceptor doped $SrTiO_3$ at room temperature (conductivity 1-4 orders of magnitude lower than in the bulk). At low temperatures the ionic conduction prevails in the grains. The existence of the grain boundary space charge layer was assigned to electronic states in the band gap formed right at the boundary itself, which is known to grow atomically sharp grain boundaries (Mecartney et al., 1980). An enormously high value for the area density of grain boundary states, assigned to donor-type grain boundary charges, was determined for $SrTiO_3$, $\sigma_{GB} = 0.56\,C/m^2$, which is about 0.5 charges per unit cell. The box shaped depletion region of charge carriers around the grain boundary extended to around 70-100 nm, while the width of the crystallographic grain boundary mismatch was on the oder of a few unit cells, or even atomically sharp (Denk et al., 1997). Vollmann et al. (1997) then developed a model, which yielded a band bending of the conduction band around the grain boundary of -2.6 eV at 300 K, which is an extremely large value. The current across the grain boundary was determined to be predominantly electronic while currents in the bulk are dominated by ionic conductivity. The local fields that will develop at the grain boundary may reach values as high as several 10^7 V/m (0.1% Fe-doping). For long times of exposure to d.c. voltages, the pile-up of defects near the grain boundaries becomes so significant that charge carrier emission sets in and a fatal breakdown current is observed. Under high fields the behavior of grain boundaries in $SrTiO_3$ stayed essentially the same as in the low field range, unless a fatal degradation current was observed (Vollmann and Waser, 1997). Despite the higher driving force and an expected enhancement of the hole (electron) current across the grain boundary, this is partially balanced by a reduced mobility of polarons under high fields. The barrier character for the oxygen vacancy motion did not change.

The crucial question to be asked at this place is whether the same effect will occur in PZT. If similar grain boundary charges are present in PZT, their effect on the ionic mobility should be similar. On the other hand the high polarizability, including the domain wall motion effects, have to be included into the considerations in PZT. Local grain boundary charges, which have a large effect on the mobility of ions in $SrTiO_3$ may be compensated by the local polarization, which reduces the effective fields to very low values. Considering the experimental observation that some of the defect agglomerates actually grow across grain boundaries, the blocking function of the grain boundaries on the ionic motion in PZT will most likely be much lower than in $SrTiO_3$. Furthermore, the leakage current does increase in PIC 151 due to fatigue(Lupascu et al., 2004), the blocking function of the grain boundary thus seems weak and most likely the grain boundary will function as the conductivity path through the sample. This also easily explains the high field and annealing data of Sect. 2.3. While the asymmetry can be easily removed due to high field treatment, the reduction in switchable polarization is extremely stable. Highly mobile charges at the grain boundary constitute the offsets and are redistributed by the high fields, while the agglomerates within the grains are highly stable entities. Even thought this looks like a beautiful picture, the role of grain boundaries is not finally answered yet.

5.8 Agglomeration

This section is devoted to the understanding of the formation of hard agglomerates as they were found in Sect. 3.1.

5.8.1 Point Defect Sinks

The easiest way to incorporate the agglomeration into the general equilibrium of point defects is to treat them as irreversible sinks for point defects, which is similar to the assumption by Brennan (1993) (Sect. 5.8.4). If the enthalpy for attaching the defect ion to the agglomerate is sufficiently high, the sink can be considered absolute. Brennan assigned a size dependent attractive force of newly arriving ions, which was neglected in the forthcoming arguments, because it would render the model too complex. For an absolute sink the rate at which ionic defects agglomerate depends on the number of starting agglomerates, which will be termed agglomeration seeds here. Similarly as in the nucleation and growth processes of phase transitions or domain switching, the number of seeds determines the initial rate of agglomeration, while the later development is determined by the growth rate of the agglomerate. The number of seeds depends on the production process of the ceramic, particularly the cooling rate, which determines the degree and the amount of frozen in defects. The starting defects are most likely larger than isolated point defects, but the model developed later does not necessarily require that.

Furthermore, the rate at which the Curie point is traversed plays a significant role in the local effectiveness of the agglomeration seed. Local stresses or stress gradients or the electric fields may become extremely high while a domain wall is passing and will attract vacancies into the agglomerate. As the microstructure determines the local stress levels, evidently the grain size distribution is also relevant.

As outlined in Allnatt and Lidiard (1993) (Chap. 3.6), the effective interaction strength between solute defects in a solid is only of short range. The driving force for reaching thermal equilibrium is high in a small volume around an existing defect, because the gradient in chemical potential is high. Thus, an agglomerate will attract all point defects around it in a certain, but small volume. The model by Brennan (1993) can be applied to this volume. Once this volume has been emptied, only mechanisms able to move larger amounts of defects will be able to bring new defects into the volume element in question. This is the major reason, why cycling is necessary to induce fatigue and why bipolar cycling is more effective. The cycling induces a fairly high ionic motion constituted by the domain wall motion-induced drift of ionic defects as already discussed in Sect. 5.4.3. Before poling inversion during bipolar cycling the additive combination of external and local fields is then a second effective mechanism capable of moving ionic defects at appreciable rates into the vicinity of the agglomerate.

The effective volume in the diffusion range of a particular agglomerate is then always refilled with new defects, until the entire volume of the grain has been completely emptied of defects. An interesting fact about the growth law of such an agglomerate is that it is independent of the dimensionality of the agglomerate. If the growth rate of a defect agglomerate is determined by the relative volume of the outer perimeter with respect to the already agglomerated ions, the dimension of the agglomerate is irrelevant. For linear chains single ions can be added to the ends of already existing chains of length l and the growth rate will be proportional to $1/l$. For a planar agglomerate the entire circumference is a growth region. The ratio of agglomerating to agglomerated ions is given by the ratio of circumference to the area of a circle, again proportional to $1/l$ and the same is true for a volumetric agglomerate. Only if the defect fluxes are unidirectional a three dimensional defect will grow differently from a two dimensional one.

5.8.2 Dislocation Loop Formation at Low Oxygen Partial Pressure

One of the most relevant publications to explain our fatigue data in PZT is the study by Suzuki et al. (2001). The search for base metal electrodes in the production of ceramic multilayer capacitors emerged into the sintering of such devices with nickel as the metal electrode material under reducing atmosphere at very low oxygen partial pressure (p_{O_2}). Ca-doped $BaTiO_3$ will develop significant oxygen vacancy concentrations if sintered under reducing atmospheres. If the ratio of the A to B-site type metal ions is set to exceed 1.00 during the powder processing, Ca-ions will partially substitute titanium forming an acceptor center. Each Ca on the "wrong" sublattice will generate one $V_O^{\bullet\bullet}$ according to

$$\mathrm{CaO} \stackrel{(\mathrm{Ba,Ca})\mathrm{TiO_3}}{\longrightarrow} \mathrm{Ca}''_{\mathrm{Ti}} + O_{\mathrm{O}} + V_{\mathrm{O}}^{\bullet\bullet} . \quad (5.17)$$

Their TEM study revealed the formation of dislocation loops in high numbers. The loops of these pure edge dislocations lay on {100} planes with Burgers-vectors <100>. Under long electron beam irradiation these dislocation loops were unstable, so the length of the Burgers vector could not be determined rotating the sample.

Despite the fact that we only found microcracks in a first attempt to investigate the microscopic effects of fatigue using TEM (see Sect. 3.7.2), I suggest that these dislocation loops are also present in PZT. I suppose that the etch grooves visible in the micrographs of Sect. 3.1 are such dislocation loops. They are necessarily charged entities due to the strong acid attack. The low stability of these dislocations loops under an electron beam correlates well with the findings of Suzuki's work. The dislocation loops there are also unstable under constant electron irradiation. Considering the fact that very low oxygen partial pressures were applied to generate these loops, the equivalent entity in PZT, which is induced due to fatigue may even be less stable than those in $BaTiO_3$. On the other hand, and this is an argument very favorable of the idea that fatigue induces the same structures in PZT, the formation of the dislocation loops under reducing conditions in $BaTiO_3$ has to occur at high temperatures, when the material is in equilibrium with the surrounding atmosphere. During fatigue the neighboring domain structure can easily stabilize the planar dislocation loops due to energetically favorable formation of 180° domains in the immediate neighborhood of the defect. This argument will be further extended later in this section. The TEM pictures of Suzuki et al. also displayed a strong interaction between 90° domain walls and these dislocation loops. The observation that the dislocation loops in $BaTiO_3$ rapidly disappear upon application of a focused electron beam suggests that the dislocation loops themselves are charged. Oxygen vacancies as a single missing layer of oxygen ions are a highly probable explanation for the observed dislocation loops and their disappearance. Oxygen is readily available in the next unit cells and the vacancies are the most mobile species

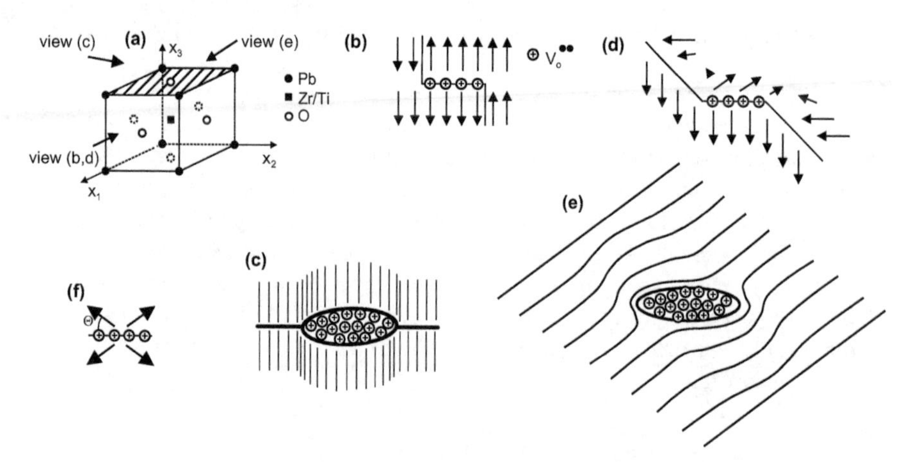

Fig. 5.3. The possible orientations of tetragonal and rhombohedral domains with respect to planar agglomerates. (**a**) The cubic perovskite type crystal structure containing a (001) agglomerate plane. The oxygen ions located above the Ti-ion are O(1) sites, located in the hatched plane, while the other two are O(2) sites (definition see Fig. 5.2). (**b**) Tetragonal 180° domain wall viewed along $[\bar{1}00]$ for a charged agglomerate plane of $V_O^{\bullet\bullet}$. (**c**) Same as (**b**) viewed at an angle from above. (**d**) 90° domain wall interacting with a planar agglomerate. (**e**) 3D-image of (**d**) depicted as lines of equal height along [001]. (**f**) Orientation mismatch ($\Theta = 35.3°$) of the rhombohedral [111], $[\bar{1}11]$, $[1\bar{1}\bar{1}]$ and $[\bar{1}\bar{1}\bar{1}]$ domains with respect to the agglomerate plane viewed along $[\bar{1}10]$

in materials, that do not contain interstitial sites for defects. Suzuki et al. then interpret their own data by the agglomeration not only of oxygen vacancies to form those dislocation loops, but to an agglomeration also of cation vacancies. These are attracted due to the electrostatic forces exerted by the doubly charged $V_O^{\bullet\bullet}$. The formation of a locally fairly complex dislocation structure is suggested. In HRTEM pictures the formation of shear structures with Burgers-vector $^1/_2$ <110> was found after long electron beam irradiation. This further corroborates the interpretation of charged $V_O^{\bullet\bullet}$ planes as the dislocation loop, which disappear upon irradiation due to their charge state and leave a structure of double barium ion or double TiO_6-octahedron layers in the dislocation loop.

Previous studies by Becerro et al. (1999) and Grenier et al. (1981) on a series of increasing concentrations of iron acceptors in the perovskite mineral $CaTiO_3$ up to the complete phase transition to $Ca_2Fe_2O_5$ showed the formation of similar planar structures as Suzuki's work. The increased amount of iron doping generates oxygen vacancies according to

$$Fe_2O_3 \overset{CaTiO_3}{\longrightarrow} 2Fe'_{Ti} + V_O^{\bullet\bullet} + 3O_O\,. \qquad (5.18)$$

The agglomeration occurs in several steps. First isolated point defects cluster to linear chains of oxygen vacancies. Once the concentration of Fe^{3+} is sufficiently high, these linear chains extend to form planar structures of tetrahedral coordination. Both their studies were largely based on Mößbauer spectra.

In another work Suzuki et al. (2000) investigated the diverse local structures induced by the strain mismatch between a $SrTiO_3$ substrate and a heteroepitaxial layer of $BaTiO_3$ grown onto it. In this case not only planar vacancy planes were found. Several different structures like $\Sigma 3$ twin boundaries were identified. It is not clear whether such energetically rich structures will also be induced in PZT during cycling due to the increasing local mechanical stresses. In polycrystalline ceramics microcracking is also a very likely mechanism to release increased stresses. These were observed in Sect. 3.7.2.

The occurrence of extremely stable planar structures was also pointed out by Smyth (1985). The perovskite structure is so stable, that it can support the incorporation or extraction of large amounts of ions without loosing its local order. The mere effect of the changes in stoichiometry are the formation of layered structures, all consistently in {100}-planes. Often entire layers of MO_6 octahedra are removed from parts of the structures, but always as entire layers.

A link to fatigue behavior of some hard defect entities was already pointed out by Warren et al. (1994a) (see also p. 21) and Scott and Dawber (2000b). Under similarly highly reducing conditions they observed that the domain switching was extremely reduced. As can be seen from the data by Suzuki et al. (2001) above, the strong reduction of $BaTiO_3$ due to H_2 will already induce highly agglomerated defect clusters.

The first-principles pseudopotential total energy calculations of Park and Chadi (1998) are also in excellent support of our findings and the model assumptions by Brennan (1993). Park and Chadi incorporated an oxygen vacancy into a 40 atom supercell and determined the nearest neighbor relaxations. Three different configurations resulted from their study. The first one is the relaxation around the vacancy V_c for a type O(1) location of the $V_O^{\bullet\bullet}$ vacancy. O(1) locations are those which form linear Ti-O-Ti chains along the c-axis, which is the polarization direction (termed site 3 by Arlt and Neumann (1988), or equivalently 1, because there is no additional acceptor in the vicinity here, see Fig. 5.2). These V_c necessarily induce tail to tail domains in their vicinity like those depicted in Figs. 5.3 (b) and 5.4 (a). The next type of location for the oxygen vacancy with respect to the polarization of the domain is O(2) (Fig. 5.3), termed V_{ab} by the authors. Two possibilities of polarization orientations exist, one which is simply the polarization orientation of the domain all around the defect V_{ab}^{sw} which is switchable (sw) and the other one is a second tail to tail configuration V_{ab}^{ud} (up-down). According to their calculations V_c is 0.3 eV more stable than any of the two V_{ab}, which they were not able to distinguish energetically. Their considerations also well explain the effect of the much less pronounced fatigue perpendicular to the

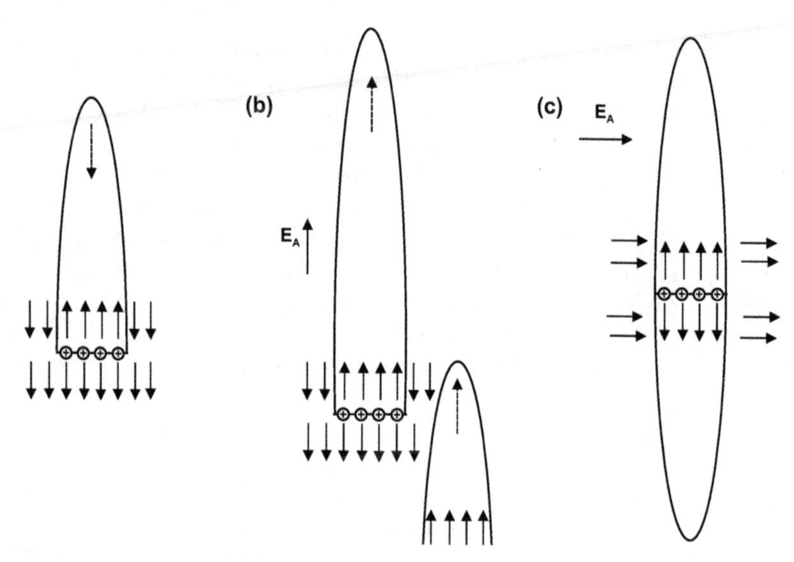

Fig. 5.4. Needle domains around an agglomerate plane: (**a**) During fatigue at a moment of applied electric field (the dashed arrow marks the movement of the domain wall), (**b**) when reverse switching occurs, and (**c**) for a hysteresis measurement perpendicular to the fatigue axis

cycling direction. In their picture all V_c turn to be V_{ab} because the coordinate system has been turned with the 90° polarization rotation. This picture only works to a certain extent for the agglomerate plane, because 90° domains form around them initially (Fig. 5.4 (b)). Nevertheless, the effect of recovery is beautifully explained even in the case of an agglomerate plane.

Another experimental result supporting the wedge shaped domain patterns as assumed by Park and Chadi (1998) was found by Kudzin et al. (1975) during fatigue measurements of [001]-oriented $BaTiO_3$ single crystals. They explicitly state the formation and growth of wedge shaped domains starting in the interior of the crystal. Initiation sites are minor defects introduced in the crystals during crystal growth (Remeika-Method). I consider these domains to be exactly the ones predicted to develop from a circular defect agglomerate. In an earlier work, the same authors had already established, that fatigue will not be induced, if liquid electrolyte electrodes were used (Kudzin and Panchenko, 1972).

A result that beautifully fits with the agglomeration of $V_O^{\bullet\bullet}$, which could not be explained previously, are the EPR results in PZT after fatigue by Warren et al. (1997c) (see p. 22). They showed that the EPR signal of the defect pair Fe^{3+}-$V_O^{\bullet\bullet}$ was changed due to the fatigue process and assigned this to a distortion of the octahedron around the Fe^{3+}-ion. This distortion can easily be provided by an agglomerate of oxygen ions. As the next neighbor interaction constitutes 70% of the EPR signal, it is predominantly the next neighbor $V_O^{\bullet\bullet}$ that determines the Fe^{3+} EPR signal, but the modifications in the proximity due to the planar defect are sufficient to act as a distortion of the local Fe^{3+} environment. As all PZT data collected by the Sandia group use thin films, the fatigue mechanism in thin films may also be the agglomeration of $V_O^{\bullet\bullet}$. This was already suggested in a recent note by Scott and Dawber (2000b) which discusses the fatigue process in a similar way as it is done here, but no direct experimental evidence was available to them. Two assumptions on the fatigue mechanism reported in the same work are somewhat critical. The first concerns their assumption, that overall the dominant pinning of domains is of electronic character not only for biasing and UV treatments, but also after fatigue. Their second assumption is that the major fatigue inducing moment during the bipolar fatigue cycle is right at the coercive field, when the domain system is more or less randomly oriented. This seems not reasonable using the acoustic emission results from Chap. 4 and the field dependent fatigue data from Chap. 2.2. According to our data it seems that only above the coercive field, when already all reordering of the domain system has taken place the major fatigue inducing effects occur.

A possible reason that fatigue is not observed in SBT ($SrBi_2Ta_2O_9$) lies in the layered structure of this material. While charge separation under UV illumination yields the same behavior as it does in PZT (Al-Shareef et al., 1997), the bipolar cycling leaves the polarization hysteresis in SBT unaffected. As SBT is a layered perovskite, oxygen deficiency will occur in planes parallel to the layer structure as is well known for the perovskite type high T_c supraconductors (Salje, 1990). As these planes are inherent in the structure, defects localized in them do not have such a significant effect on the switchability of domains as they do when forming extra planes in PZT interacting with 90° domain walls (see Fig. 5.3).

Another observation by the Sandia group showed that a polarization reduction induced by bipolar electric fatigue (70% loss in polarization) can only be recovered to a small degree (by about 20%) using UV light and a bias field applied to $BaTiO_3$ crystals (Warren et al., 1995b). Thus the structures induced by bipolar fatigue are definitely of ionic character. Their stability indicates that agglomerates may be involved also in the case of $BaTiO_3$.

A question arising from the different etch groove pictures in Figs. 3.1 and 3.2 is whether such structures can actually cross grain boundaries. This may be answered by the simple observation that domain structures communicate with each other across grain boundaries. This is evident from SEM pictures like those shown for the domain structure by Arlt (1990b) or the direct observation of domains crossing grain boundaries by atomic force microscopy (AFM) (Gruverman et al., 1996b, 1998b,a). The images in Figs. 3.1 and 3.2 convey the same message demonstrating that the agglomerates will grow across the grain boundaries.

The formation of stable fixed polarization was also observed using AFM. Gruverman et al. (1996b, 1998b) showed that islands of non-switching polar-

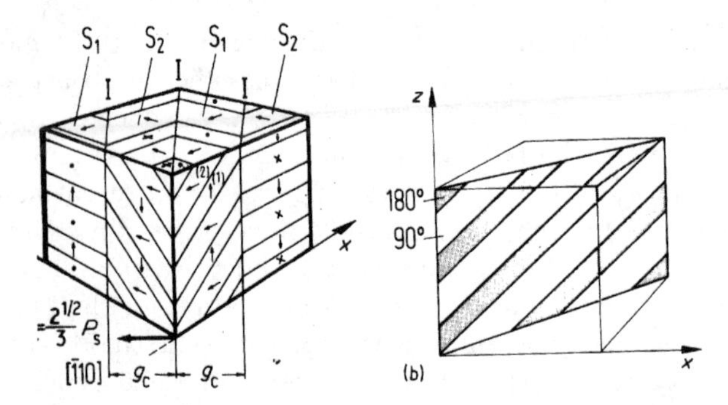

Fig. 5.5. Possible orientations of tetragonal domains filling space without free charges present (Arlt (1990b), ®Kluwer Academic Publishers). (**a**) Three dimensional picture. (**b**) Plane between striped domain wall structures containing 90° and 180° domain walls itself

ization develop during fatigue. These show fractal contours and grow with increasing number of cycles.

Figure 5.5 displays the well known 3D patterns of domains, that can develop in tetragonal $BaTiO_3$. 24 different patterns were determined to possibly exist by Arlt and Sasko (1980), called α-patterns, each containing not a single charged domain wall. The development of a charged domain wall in such a structure will immediately disturb it and generate other charged domains. On the other hand, a different spatial domain arrangement is frequently found in $BaTiO_3$, called β-pattern, which inherently contains charged domain walls. The typical property of β-type patterns are alternating wide and narrow domains and alternating 90°, 0° and head to side boundaries, the latter being charged. If more than one of these structures occurs in one grain, more complicating boundaries develop, which were not investigated. Even though charged areas exist in this structure, they are parts of a {011}-plane. They cannot be constituted by the agglomerates which are {001}-planes. Thus in all cases, the planar defect agglomerates will disturb the existing domain pattern and generate wedge shaped domains of their own. This is one further reason, why the defect agglomerates significantly suppress the domain wall motion. In PZT the herringbone patterns are not as frequent, but nevertheless commonly exist in larger PZT grains. Thus the same type of 90° and 180° 3D-structure exists as in $BaTiO_3$.

5.8.3 Induced Crystallographic Phase Change

As the agglomerates occupy well defined planes in the cubic prototype crystal structure the forces that these planes will exert on certain polarization directions will be different in the tetragonal and rhombohedral crystal structure. The shear stresses exerted on domains in the tetragonal phase vanish for certain domains, while in the rhombohedral crystal structure every domain experiences some shear stress. Thus phase changes may be induced by the defects as was demonstrated in Sect. 3.3.

An indication of enhanced tetragonality after fatigue was found in Raman spectra. Güttler et al. (1995) report the mode hardening of the Raman spectra during fatigue. This development is in the similar direction of development as the crystallographic phase change towards a more tetragonal structure in PIC 151. I consider it possible that planar agglomerates also occur in the 300 nm thick films (PZT 48/52) that were investigated by Güttler et al. (1995), because a tetragonal structure seems to also be favored by the fatigue process even in the case of thin films. Simply no chemical etching studies have been performed on thin films yet.

A similar observation was also reported on X-ray data. Pan et al. (1992a) derived an increased tetragonality of fatigued PLZT 7/58/42 and 7/60/40, which are compositions near the morphotropic phase boundary at that high level of La-doping, from a slightly more pronounced valley between the two {200} and {002} peaks of the tetragonal phase after fatigue. This valley is the location of the rhombohedral {002} peak. This analysis needs exquisite fitting to be meaningful, but the authors considered it to be sufficient. In a second work the same authors (Pan et al., 1992c) extended their analysis and compared different degrees of texturing of PZT due to poling, thermal depoling, a.c. depoling, and fatigue. They were again able to show that bipolar fatigue cycling yielded a much better separation of the tetragonal (200) and (002) peaks.

An interesting observation in the context of crystal structure was the analysis by Al-Shareef et al. (1996a) on thin film $RuO_2/Pb(Zr_xTi_{1-x})O_3/RuO_2$. They found that despite the RuO_2-electrodes their thin films (400 nm thick) degraded after about 10^5 cycles if highly tetragonal compositions were chosen. They used arguments along the lines of defect chemistry to explain their data due to only a limited solubility of oxygen in RuO_2. The electrode can thus only function as an oxygen sink to a limited extent. The higher content of $V_O^{\bullet\bullet}$ in Ti-rich PZT compositions (Prokopalo, 1979) further supports this scenario. They also observed that the morphology changed from columnar to granular across the film thickness. The authors themselves ruled out several other reasons for the stronger fatigue in Ti-rich PZT and ended up with the change in morphology as one major cause, which had also been reported by other authors (Klee et al., 1993). This observation along with our own data suggests, that once a grain structure contains grain boundaries effectively parallel to the external electrodes, it to a certain extent confines the oxygen vacancies to their grain of origin. Then similar agglomeration processes may develop as in the bulk case of PIC 151 above. Another fact which the authors did not discuss in detail is the higher tetragonality of the Ti-rich PZT compositions and the inherently higher microstresses. As the films are very thin, microcracking will most likely not take place and has so far not been observed, but the higher transverse stresses may just add to the electric field effect and enhance the fatigue effect, which would be equivalently obtained for a higher cycling voltage.

5.8.4 Iterative Models in General

The iterative character of fatigue has been previously addressed by Yoo and Desu (1992c,a, 1993); Yoo et al. (1993); Brennan (1993); Desu and Yoo (1993a) and Shur et al. (2000). In view of the agglomeration encountered in Chap. 3 and its discussion in the preceding sections it has become clear that fatigue in PZT is at least in large parts an iterative process of defect agglomerate growth.

The Model by Brennan

The first step towards the understanding of fatigue due to hard agglomerates was taken by Brennan (1993). He modelled the fatigue to be due to the agglomeration of charged defects localized next to previous defects and forming planar agglomerates of the form depicted in Fig. 5.3 (b), but did not specify a crystallographic axis to it. He also assumed vacancies to be the relevant defect type. His uniaxial model was based on a Landau-type expression of the free energy F of the domain with an additional term including a charge density on a charged plane of point defects. This term is a direct result of Maxwell's law div$\mathbf{D} = \rho_{free}$, due to the free charge of the defect.

In order to arrive at the logarithmic dependence with cycle number, he assumed that the activation energy of attaching a new vacancy to the already existing agglomerate is proportional to the size of the already existing agglomerate. This assumption directly yields the iterative character of the model and the logarithmic dependence. This is not exactly the assumption taken for the model in this work where a size independent capture is assumed (see beneath). He arrives at the following expression for polarization:

$$P - P_o = \frac{kT}{c} \log(N) \tag{5.19}$$

with a proportionality constant c.

A direct result of the Landau-type approach is that the ferroelectric hysteresis $E(P) \approx \alpha P + \beta P^3$ will be modified. Brennan assumed the material to be constituted of the two opposing domains of equal size only and added the two fields of the domains, one adding the polarization due to the charges and one subtracting it. This yields a decreasing remnant polarization as expected, but also a decreasing coercive field, which is evidently contradicting experimental data. Considering that his model does not include any 90° domain effects, it already contains the essentials of the agglomeration effect.

The Model by Yoo and Desu

In contrast to Brennan, who did not assume any location of the agglomeration within the specimen, Yoo and Desu (1992c,a,b) considered point defects to be agglomerating next to each other underneath the electrodes of thin films or at other internal boundaries (Yoo et al., 1993). On the other hand, they did not assign a particular geometrical structure to such agglomerates. Their results are supported by the high influence of metal and oxide electrodes on fatigue. The interpretation of agglomeration was, that a particular defect will hinder domain wall motion, nucleation, or growth to a certain extent, but that the relative effectiveness of a larger obstacle would not scale with size but be less pronounced. Thus an incremental growth of the obstacles was assumed and a decreasing influence on the switching process per increment.

Due to the general implications of their model on any iterative model of fatigue the lines of thought will be reproduced here.

The movement of oxygen vacancies is assumed to be dominated by the local field. This in turn is determined by the external field plus the locally generated fields due to the spontaneous polarization and the free charges present. The local field induces an ionic current of defect ions, here oxygen vacancies, according to standard Ohm's law with thermally activated ion jumps (Hench and West, 1990)

$$j = v\rho \mathrm{e}^{\Delta S} \mathrm{e}^{-Q/kT} \mathrm{e}^{zqbE_L/2kT} \quad . \tag{5.20}$$

Their first argument that agglomeration occurs due to the structural instability of grain boundaries and electrode interfaces is only true, if they consider it to be the room temperature equilibrium. During sintering the times to equilibrate the defects in grains and grain boundaries are sufficient. Only in the case of equilibrium with the later applied electrodes a non-equilibrium defect population is frozen in. Particularly a significant difference in equilibrium concentration at grain boundaries between the freezing temperature of oxygen vacancies at about 700K and room temperature with respect to the grain solubility is not expected.

The agglomeration rate per cycle is set proportional to the effective field difference between forward and reverse conductivity, which is a reasonable assumption.

The next assumption in their derivation is that the rate of change with respect to the existing accumulated charge is proportional to the presently existing field difference, which yields an exponential law for the rate of decrease of the field difference.

$$\frac{d\Delta E_i}{dN} = -\xi \Delta E_i \quad . \tag{5.21}$$

The next assumption in their work is that the change of polarization with respect to fixed charge is proportional to the polarization itself.

$$\frac{dP}{dN} = -\nu P. \tag{5.22}$$

They assume that there is a higher probability of loosing switchable polarization, if more polarization is present.

Equations 5.20 to 5.22 lead to a power law loss of polarization (Yoo and Desu, 1992a):

$$P = P_o(An + 1)^{-m} \tag{5.23}$$

with n the cycle number, A a piling constant and m the decay constant.

As the oxygen vacancies agglomerate, the effective field is reduced. The effect of additional defects is less pronounced, because they only alter an already existing agglomerate. Due to the agglomeration the defects induce an additional local field which reduces the local field. At this point there is a little subtlety in their argument, which will probably not alter the final result, if just the mechanism is corrected, but is not correct in itself. If the electric field is diminished for one polarity, it will be enhanced for the opposite one. As the thin film also has two electrodes, the accumulation should equally well occur on both sides and the effective field due to the charges would vanish. The incremental reduction of the alternating local field amplitude ΔE is thus not correct. But if one assumes that the domain wall velocity is reduced due to agglomerates by just enhancing the time of fixation at that location, a similar reduction in domain wall velocity is achieved.

They also assume the bulk concentrations of defects to be large and not be altered by the fatigue process, which is different from the assumption here, that the total number of defects is fairly low and their total number fixed.

The Model by Dawber and Scott

Dawber and Scott (2000a,b) use the same ionic current density difference to calculate the growth of a space charge layer. Different from Yoo and Desu (1992c,a) they do not assume that the point charges accumulate near the electrodes. Any internal diffusion layer is sufficient to stop the defect motion and generate an area of ionic space charge. They integrate the influx of charged ions to obtain the change of potential drop V over the space charge layer with cycle number n. The major assumption not immediately evident is that $P = A \cdot V_o(n = 0)/V_o(n) + B$. The appealing result of their model is that almost all experimental data are reproduced only using elementary data on PZT, including temperature dependence of fatigue rate, maximum field and frequency dependence.

The Model by Shur

An iterative model for local screening was developed by Shur et al. (2000, 2001a,b). It is based on the general description of switching in ferroelectrics as a nucleation and growth process analogous to phase transformation theory (Kolmogorov, 1937; Avrami, 1939, 1940, 1941; Ishibashi and Takagi, 1971; Ishibashi, 1985). The starting configuration is a fully poled uniaxial single crystal ferroelectric, which can be mapped to a 2D model. During switching

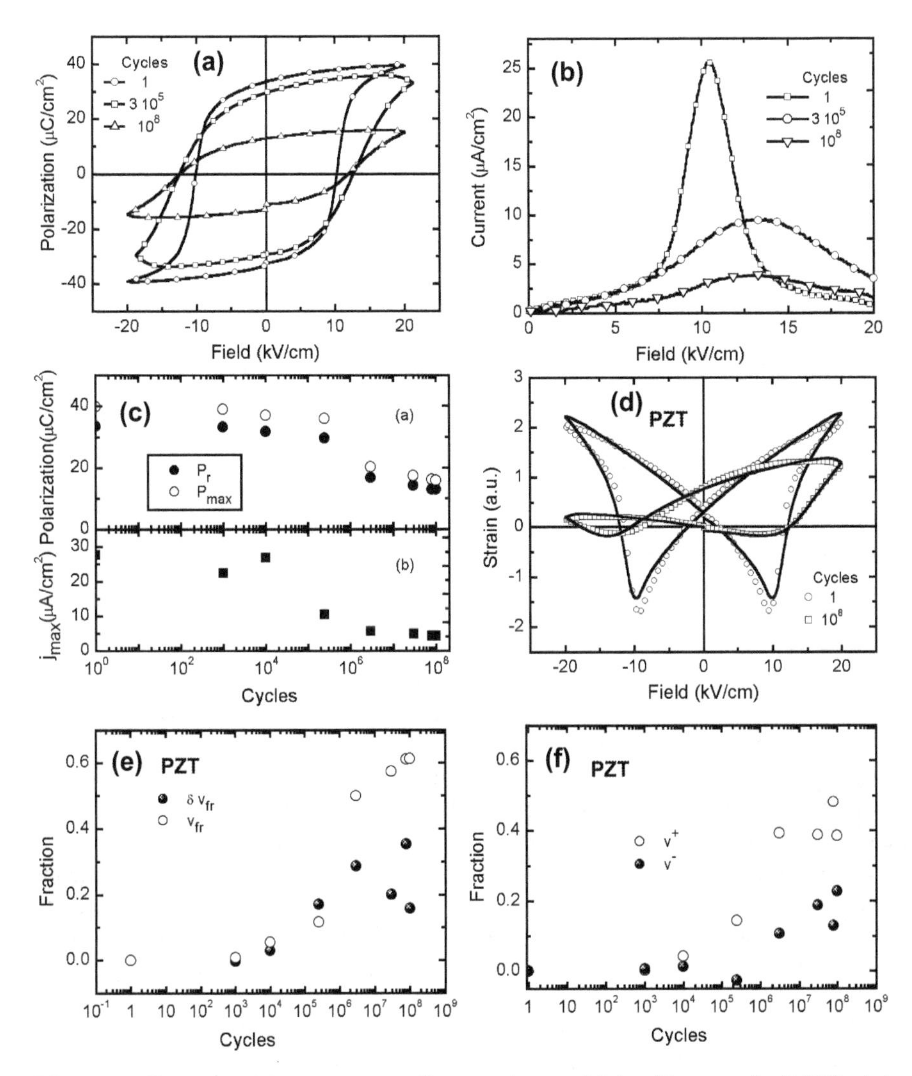

Fig. 5.6. Fit to our data set according to the model by Shur et al. (2000). (**a**) Polarization hysteresis and fitting points for three cycle numbers, (**b**) corresponding switching current data, (**c**) dependence of maximum and remnant polarization on cycle number, (**d**) strain hysteresis and fitting points for 0 and 10^8 cycles, (**e**) strain fitting parameters $\delta \mathrm{v_{fr}}$ and $\mathrm{v_{fr}}$, (5.26), and (**f**) fractions of frozen domains of each polarity

two processes simultaneously occur, the nucleation of new domains and the subsequent growth of these domains. The nucleation probability is given by

$$p\left(E_{loc}(x,y,n_f)\right) = \exp\left(\frac{-E_{ac_n}}{E_{ex} + E_b(x,y,n_f-1)}\right) \tag{5.24}$$

with a certain activation field E_{ac} needed for the onset of nucleation. E_b is the local bias field due to certain defects or clamped domains. The model itself does not determine any microscopic origin of this local bias, but allows any type of it to iteratively change in each step of cycling. Once the domain is nucleated, it will grow at two different velocities across the plane, one for plane fronts (wall motion) and one for edges. The two velocities are again determined by an activation electric field, E_{ac_w} for wall movement: $v_w\left(E_{loc}(x,y,n_f)\right) = v_{wo}\exp\left(-\frac{E_{ac_w}}{E_{ex}+E_b(x,y,n_f-1)}\right)$ and E_{ac_s} for step growth velocity: $v_s\left(E_{loc}(x,y,n_f)\right) = v_{so}\exp\left(-\frac{E_{ac_s}}{E_{ex}+E_b(x,y,n_f-1)}\right)$. The crucial parameter that makes this model different from the previously discussed models is the fact that different time dependencies enter. Shur et al. introduce the screening time τ. This is the time that local charge carriers need to rearrange and compensate the uncompensated local remnants of the depolarizing field. The fatigue rate now becomes correlated to the cycling frequency and the relaxation time τ (screening time). The local bias field changes due to τ and the switching cycle period T:

$$E_b(x,y,n_f) = E_b(x,y,n_f-1)\exp\left(\frac{-T}{\tau}\right) + E_{rd}\left(1-\exp\left(\frac{-T}{\tau}\right)\right)\left(\frac{\Delta T(x,y,n_f)}{T}\right). \quad (5.25)$$

Beyond certain cycle numbers a large fraction ν_{frozen} of the plane experiences local bias fields sufficient to suppress the switching. When these inactive regions take up large areas all switching has to occur as domain wall motion starting from the clamped crystal area. The growth process reduces from two-dimensional to one dimensional. Shur terms this transition the geometrical catastrophe. As the process itself is steady, the transition between both current types is smooth and fractal in character, which is not incorporated into the model.

The volume fractions $\mathrm{v_{fr}}$ of frozen domains are given for each polarity $\mathrm{v_{fr}^+}$ and $\mathrm{v_{fr}^-}$, and the offset polarization $\delta\mathrm{v_{fr}}$ by

$$\delta\mathrm{v_{fr}} = \mathrm{v_{fr}^+} - \mathrm{v_{fr}^-} \quad (5.26)$$

$$\mathrm{v_{fr}}(N) = \mathrm{v_{fr}^+} + \mathrm{v_{fr}^-} = \frac{P_r(1) - P_r(N)}{P_s}$$

with N the cycle number. The model well matches the images presented by Gruverman et al. (1996a) on the fractal islands of fixed domains in thin films. Due to the constants the model is very versatile. It can perfectly fit our set of switching data, the offset-polarization, the switching current data and the loss in switchable polarization as depicted in Fig. 5.6.

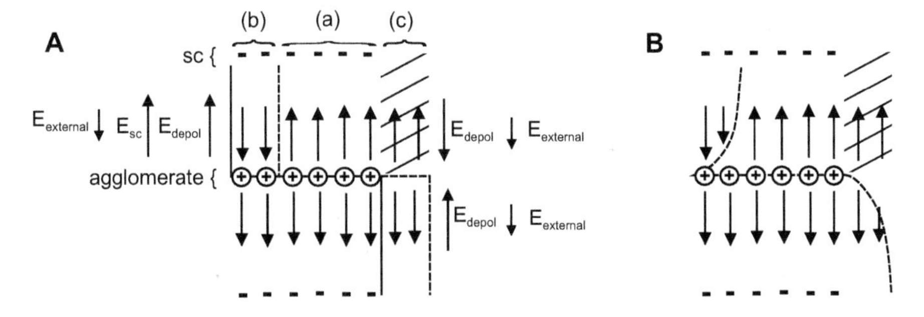

Fig. 5.7. Stepped 180° domain wall fixed at an agglomerate under external field. The external, depolarizing and space charge fields are shown and explained in the text, sc = space charge

5.8.5 Iterative Model of Bulk Agglomeration

Structure of the Model

The following model is conceived for the description of fatigue due to hard agglomerates. It does not consider the effect of electronic space charge as a primary part of the fatal fatigue mechanism, even though many of the effects encountered experimentally are due to space charges as discussed in the previous sections. Both Yoo and Desu (1992c,a) and Dawber and Scott (2000b) describe models based on the assumption, that space charge layers develop underneath the electrodes of ferroelectric thin films and constitute the major agglomeration location of charges limiting the switching of polarization. The following model considers planar agglomerates in the bulk to be responsible for the dominant fatigue effect.

Fields Around the Agglomerate

The first question is the movement of charged ionic defects in bulk grains. The previous models have considered the motion of charged defects under an applied electric field E_A and additionally an internal field E_i (Yoo and Desu, 1992c) or a space charge field E_{sc} (Dawber and Scott, 2000a) using their notations. In both cases the space charge/internal field is considered small and ion motion is dominated by the external field. This is the first fact treated differently here.

First, a stable position of an essentially planar domain wall around an existing agglomerate is considered (continuous line in Fig. 5.7, which is equivalent to Fig. 5.3(b) and (c)). An external field in favor of the left domain will try to move the domain wall to the right (dashed line). Before the domain wall is able to break loose from the agglomerate, it will bend out towards its desired position as illustrated in Fig. 5.7 B. For a schematic illustration of the different effects, the bending is omitted in part A. In section (a) of the agglomerate any ions in the bulk volume only experience the external field. The field due to the spontaneous polarization is entirely screened by the oxygen vacancies (marked ⊕) and the electronic space charges (marked -) constituted by small polarons or other localized electronic states in not too large a distance from the agglomerate.

In the lower part of section (b) the effective field on ions is simply the external field, like in section (a) and is directed away from the agglomerate. In the upper part of section (b) the effective field is also directed away from the agglomerate and it is large, because the screening and dȇpolarizing fields add up. The drift of ions will be in upward direction despite the external field in the opposite direction. Those oxygen vacancies in section (b) which are part of the agglomerate itself, on the other hand, only experience the external field plus the depolarizing field, because they are constituting the screening field themselves. As the agglomerate itself exerts a certain binding force on its constituents, these ions are considered fixed, which is equivalent to the assumption of an absolute vacancy sink.

In section (c) the most relevant things happen. Here no space charges are present to screen the depolarizing field due to the spontaneous polarization. In the upper part (hatched) the external electric field even adds to the depolarizing field. As the ion current exponentially depends on the sum of all fields, this moment yields the highest ion fluxes, which is the essential point made by Sturman (1982). The slow fatigue in thin films observed by Colla et al. (1998b) can be interpreted along the same lines. They did a set of measurements, where they cycled their films at 1.7 mHz. When they directly applied the full voltage of 2 E_c with rectangular pulses, no significant fatigue was observed. When they applied only voltages just beneath the coercive field and then only briefly raised the voltage to 2 E_c within the same cycle (again rectangular), the material fatigued within 20 cycles. Along with this cycling procedure the voltage dependent permittivity considerably dropped at E_c, while it slightly increased for high d.c. bias fields. These latter data are similar to results found by Mihara et al. (1994a). Also in thin films it seems that the ion movement is very strong under conditions, where the external field is opposing the polarization direction and external and depolarizing fields add up. Returning to our picture of this process, in the lower part of section (c) the effective field is still towards the defect agglomerate. Any ion moving towards the agglomerate in the hatched region thus becomes definitely trapped.

To get an impression of the electric fields involved, the maximum external field of 2 kV/mm (2 E_c) has to be compared to the local fields. In the case of no free charges present, $D = 0$ (in the hatched region), and without external field, the electric field is given by the spontaneous polarization: $E_{depol} = \frac{P_S}{\epsilon_o \epsilon_r}$. For $PbTiO_3$, $P_S = 0.52\,\mathrm{C/m^2}$, $\epsilon_r = 200$, this yields $E_{depol} = 300\,\mathrm{kV/mm}$. In PZT the effective dielectric constant may be somewhat higher, but one may not use the low frequency dielectric constant here, because the single domain is a monocrystal and domain wall motion does not contribute to ϵ_r. Thus, already a small fraction of the depolarizing field will by far exceed the external field. The crucial relevance of fatiguing fields (anti-)parallel to the polarization axis was directly observed by Ozgul et al. (2001) (see Sect. 5.9 beneath).

Another important indication, that the orientation of the spontaneous polarization in PZT, and thus the depolarizing field, is relevant for the enhanced fatigue behavior was given by Pan et al. (1989). Antiferroelectric compositions show much less degradation than similar ferroelectric compositions in the system $(Pb_{0.97}La_{0.02})(Zr_xTi_ySn_z)$.

If the agglomerate is not located at a domain boundary traversing larger parts of the grain, but within an existing domain, it will develop a domain wall of its own. Considering the agglomerate to be essentially circular, once it has grown to more than a few unit cells, it will induce a tail-to-tail structure around it. If the surrounding domain shares one of these orientations, a single wedge shaped domain develops (Fig. 5.7(a)). Kudzin and Panchenko (1972) observed such structures during the bipolar fatigue of $BaTiO_3$. Figure (5.8(a)) shows such a wedge shaped domain at its agglomerate. It obviously reaches into a domain of opposite polarization. Thus, the domain wall is charged on the upper part and will attract negative point defects yielding a space charge. Under the application of an external field the domain will shrink (Fig. 5.8(b)). The external field starts acting on the space charges, which will start to migrate away from their original positions. The interesting question now arises to which field those oxygen vacancies experience, which are not yet part of the agglomerate. The fields above the agglomerate do not get altered in a significant manner and essentially stay in the configuration of section (a) in Fig. 5.7. Not much ion movement will result.

The situation completely changes, when the external field direction is reversed. Before a remotely located domain wall will approach the defect agglomerate like in Fig. 5.4(b), a situation like in Fig. 5.8(c) arises. Now we have the same case as in Fig. 5.7, section (c). The external electric field adds to the depolarizing field. Charges, which are located at remote distances to stabilize the larger outside domain, are not so effective, particularly, when the domain system is essentially compensated like in Fig. 5.5 and simply no external screening charges exist. The ion flux is huge towards the agglomerate, particularly at its rim, where the agglomerate itself does not exert any repulsive force on the incoming $V_O^{\bullet\bullet}$. The incoming $V_O^{\bullet\bullet}$, when attached to the agglomerate rim will even aid the growth of the upper domain favored by the external field.

For simplicity, it is assumed that all ions entering a cylindrical volume beneath the agglomerate are captured into the agglomerate plane. In order to reduce the algebraic calculations while maintaining a reasonable physical picture the following assumptions were chosen.

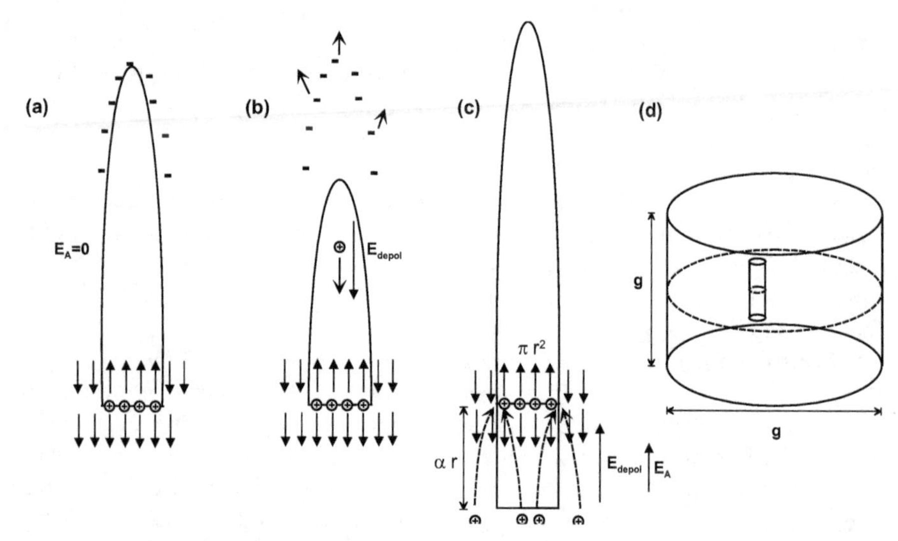

Fig. 5.8. Assumptions for the agglomeration model. (**a**) Wedge domain as stabilized by an existing agglomerate (+). The wedge itself is a charged 180° domain wall, equilibrated by electronic space charges (−). (**b**) The wedged domain shrinks due to the external field leaving the space charges behind. (**c**) The depolarizing field and the external field add up. A net current of $V_O^{\bullet\bullet}$ crosses the area πr^2 into the cylinder of height αr and volume V_A. Incoming $V_O^{\bullet\bullet}$ accumulate at the edge of the agglomerate. (**d**) The agglomerate and the cylinder of vacancy capture in the cylindrical grain

Model Assumptions

- The agglomerating species are oxygen vacancies (the model is actually not limited by the sign, so other ions may do the same).
- The capture of ions into the agglomerate is irreversible.
- The agglomerates lie on (001) planes in the tetragonal unit cells.
- The major influx of vacancies to the agglomerate is along the field direction, particularly, when the field is opposing the direction of polarization, i.e. when both fields add up (Fig. 5.8).
- The agglomerate induces a stable domain in its vicinity with the tail of the polarization vectors pointing towards the agglomerate (tips for the other sign).
- Oxygen vacancies within a certain volume beneath the agglomerate are drawn to the agglomerate under reverse field (Fig. 5.8(c))
- All oxygen vacancies are confined to their grain of origin at the cycling temperatures considered.

The last assumption is actually not really bound to the grain structure. In the case of acceptor doping, the grain boundary is an invincible barrier to positively charged ions or to holes (Waser and Hagenbeck, 2000). The same is true for negative charge carriers in the case of excess donor doping. In the model the restriction is more to be understood in the sense that the total volume of the sample contains enough primary defects as agglomeration sources, which each empty a volume around them of defects. The assumption then reduces to there being one such center per grain. Some agglomerates will actually grow across grain boundaries as seen in Fig. 3.5, particularly (b) and (d), so the grain boundary. If the grain boundary is assumed a barrier to defect diffusion the growth of the agglomerates across the grain boundary has to be carried by electric fields crossing the boundary. In PIC 151 the grain boundaries are not considered invincibly high, so the limitation that defect diffusion does not to take place across grain boundaries while there is only one starting seed for agglomerate growth is actually another way of limiting the diffusion volume of incoming defects to the agglomerate. Reversing the interpretation, one assumes a certain density of agglomeration seeds. The grains are also considered cylinders (Fig. 5.8(d)) for convenience, but this does not enter the calculations.

Calculation

During one bipolar electric cycle (of number n) the electric field is directed in its most effective way with respect to the already existing domain around the agglomerate, when it opposes the spontaneous polarization of the domain (Fig. 5.8(c)). Both the external and the depolarizing field add up and drive the $V_O^{\bullet\bullet}$ so close to the existing agglomerate into the cylinder of height αr and volume

$$V_A = \alpha\pi r^3 = \frac{\alpha}{\sqrt{\pi}} N_A^{\frac{3}{2}} b^3 \tag{5.27}$$

that they are captured and localized at the edge, with N_A the number of agglomerated ions and b the lattice constant. This only occurs during a fraction ζ of the period $\Delta t = 1/f$ of the external electric field applied at the frequency f. It is already clear at this point, that the absolute time of residence just beneath the coercive field is the relevant parameter. The frequency indirectly enters. Only during a fraction of the total bipolar cycle the local fields are favorable for maximum ionic flow. The ionic flow is given by the total time at high field.

The agglomerate itself is considered to be densely packed. Thus for every ion incorporated, the area of the agglomerate will increase by b^2 and the radius of the agglomerate is related to the number of agglomerated ions by

$$\pi r^2 = N_A b^2 \ , \tag{5.28}$$

which already entered (5.27).

The material was sintered in some oxidizing or reducing environment, which determined the total concentration ρ_o of $V_O^{\bullet\bullet}$ at the beginning of the cycling. No further exchange of ions with the outside will take place in a bulk sample or in a metal electroded thin film. The ion flux under an external applied electric field is given by (Kingery et al., 1976; Hench and West, 1990; Yoo and Desu, 1992c):

$$j_e = 2\rho ezbP \sinh\left(\frac{Eeb}{2kT}\right) \tag{5.29}$$

$$P = \xi\frac{kT}{h}\exp\left(-\frac{G}{kT}\right), \tag{5.30}$$

j_e = charge current area density, ρ = the volume density of charged defects, e = elementary charge, z = the charge number of the defect, b = hopping distance for the ion (vacancy) between two sites of residence = lattice constant, P = thermally activated probability of an ion jump to the next lattice site, including the attempt frequency $\xi\frac{kT}{h}$ (see Sect. 5.4.2) and the energy barrier without an external field present, E = electric field, k = Boltzmann's constant, ξ = a coordination parameter of the defect lattice site, h = Planck's constant, and G = the Gibbs free energy for the defect migration. γ is introduced to aid the later calculations. The particle flux $j = j_e/ze$ becomes

$$j = \rho\gamma. \tag{5.31}$$

Two sources of ion flux into the capturing cylinder can take place, the ion flux due to the electric fields:

$$\left.\frac{dN_A}{dn}\right|_j = -j(n)\Delta t\pi r^2(n) \tag{5.32}$$

$$= \rho_g(n)\gamma\pi r^2(n)\text{with}$$

$$\gamma = -2\zeta\Delta tbP\sinh\left(\frac{Eeb}{2kT}\right) \tag{5.33}$$

and the ion flux due to the growth of the agglomerate itself into the volume of the surrounding grain:

$$\left.\frac{dN_A}{dn}\right|_{\text{Vol}} = \rho_g(n)\frac{dV_A}{dn}. \tag{5.34}$$

For the latter the assumption is that the aspect ratio of the capturing cylinder does not change depending on its size. The minus sign in equation (5.32) is due to the fact that the surface has an outward normal, while the particle flux is into the volume.

The total flux of ions into the volume V_A becomes:

$$\frac{dN_A}{dn} = \left.\frac{dN_A}{dn}\right|_j + \left.\frac{dN_A}{dn}\right|_{\text{Vol}}. \tag{5.35}$$

As will be seen later, the total absolute number of defects N enters the calculations through the predetermined density $\rho(n=0) = \rho_o$ and the confinement to the grain of size $\pi g^3/4$. Thus it is convenient to introduce the density of defects in the grain $\rho_g(n)$, which have not yet been captured into the agglomerate:

$$\rho_g(n) = \frac{N - N_A(n)}{V - V_A(n)}. \tag{5.36}$$

Using equation (5.27) one obtains:

$$\frac{dV_A}{dn} = \frac{3\xi b^3}{2\sqrt{\pi}} \sqrt{N_A} \frac{dN_A}{dn} . \tag{5.37}$$

Using (5.36) and (5.37), and rearranging the terms in equation (5.35):

$$\frac{dN_A}{dn} \cdot \left(1 - \sqrt{N_A} \frac{3\alpha b^3}{2\sqrt{\pi}} \frac{N - N_A}{V - V_A}\right) = \gamma N_A b^2 \frac{N - N_A}{V - V_A} \tag{5.38}$$

one obtains the integral equation:

$$\int_1^{N_A} \frac{V - V_A - \frac{3\alpha b^3}{2\sqrt{\pi}} \sqrt{N_A}\,(N - N_A)}{\gamma N_A b^2 (N - N_A)} dN_A = \int_0^n dn . \tag{5.39}$$

The integration limits are chosen such that one starts with one initial defect, which is also necessary mathematically later, and zero cycles. The solution to equation (5.39) is straight forward and given in the appendix. Unfortunately the expression does not permit an easy rearrangement in terms of n. The implicit solution is given by:

$$n \approx -\frac{1}{\gamma b^2 \rho_o} \ln \frac{N - N_A}{N_A} + \frac{3\alpha b}{2\sqrt{\pi}\gamma} N \ln \frac{\sqrt{N} - \sqrt{N_A}}{\sqrt{N} + \sqrt{N_A}} - 2\frac{\alpha b}{\sqrt{\pi}\gamma} \sqrt{N_A} . \tag{5.40}$$

For the experimental conditions in our experiments the agglomerate growth is plotted in Figs. 5.9 through 5.13. The parameters are $\xi = 1$, $\alpha = 1$, temperature $T = 350$ K (in our experiments), the fraction of active cycle $\zeta = 0.5$, grain size $g = 5\,\mu\text{m}$, $G = 1\,\text{eV}$, $\rho_o = 10^{-3} \cdot b^3$.

A fairly beautiful support of the formation of needle domains are the strong modifications of the domain patterns observed after fatigue. Figure 5.14 shows atomic force microscopic images before and after fatigue (Lupascu and Rabe, 2002). A complete breakup of the initially watermark domain pattern into a system of narrow needle domains is observed. Such a breakup of domain systems into smaller units has been theoretically predicted due to a dead layer in thin film ferroelectrics (Bratkovsky and Levanyuk, 2000a). As this domain pattern was observed throughout the entire 1 mm thick sample, a direct influence of dead layers underneath the electrodes can be excluded for the bulk studies here, but in a similar sense, dead layers may form at the perimeter of the crystal, namely the grain boundary in a ceramic.

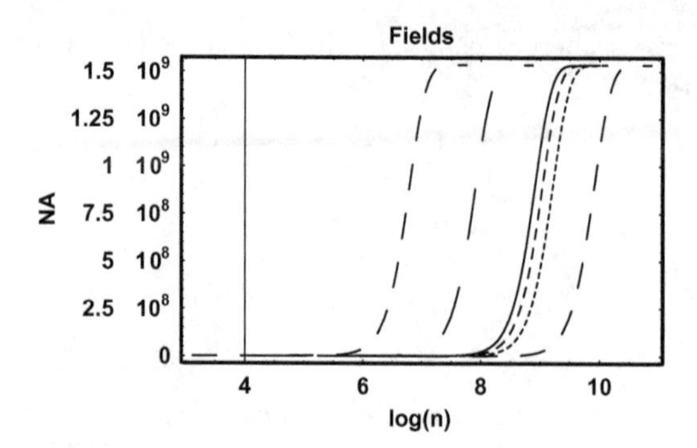

Fig. 5.9. Field dependence of agglomeration according to equation (5.40) for the local fields (left to right): 200 kV/mm, 20 kV/mm, 2 kV/mm, 1.5 kV/mm, 1 kV/mm, and 0.2 kV/mm

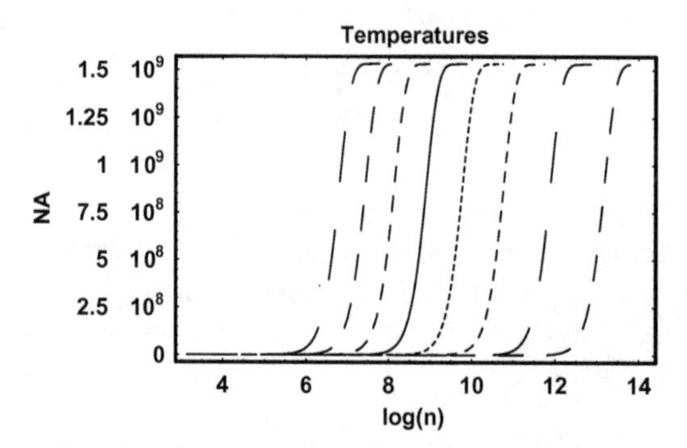

Fig. 5.10. Temperature dependence of agglomeration according to (5.40) for (left to right): 410 K, 390 K, 370 K, 350 K, 330 K, 310 K, 290 K, 270 K

Discussion of the Model

The temperature dependent data mentioned in Sect. 1.12 contradict the findings of the model, meaning that their studies showed a lower fatigue at elevated temperatures in $BaTiO_3$ at 80°C than at 25°C (Kudzin et al., 1975). On the other hand, Jiang et al. (1994d) reported a strictly fatigue free behavior of PLZT 9.5/65/35 at -140°C, when it exhibits a widely open hysteresis loop. Thus, two competing mechanisms may play a role here, the less open hysteresis loop at higher temperatures, which also means lower depolarizing fields, and the enhanced mobility of the ions, which should actually accelerate the fatigue.

The temperature dependence (Paton et al., 1997) of the fatigue rate on the other hand perfectly fits the model just developed. These authors showed that the fatigue rate at 100 K is completely negligible, while at 500 K a very rapid decay of polarization was observed in their material (see also p. 34).

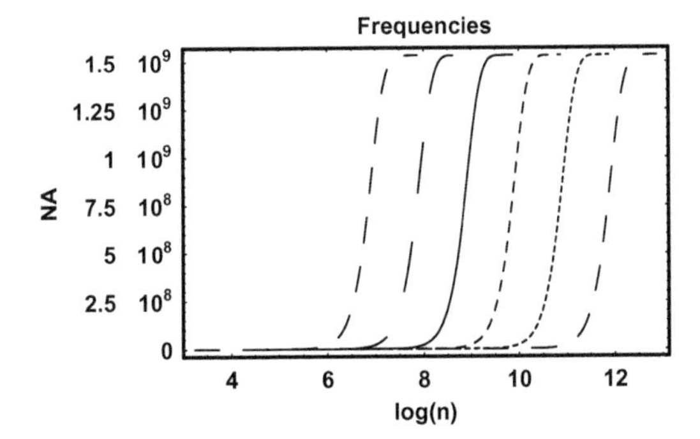

Fig. 5.11. Frequency dependence of agglomeration according to (5.40) for (left to right): 0.5 Hz, 5 Hz, 50 Hz, 500 Hz, 5 kHz, 50 kHz

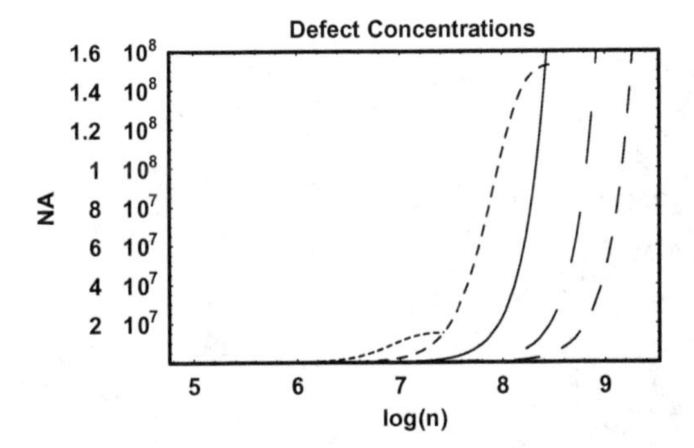

Fig. 5.12. Dependence of agglomeration according to (5.40) depending on the initial defect concentration (left to right): $5 \cdot 10^{-2}$, 10^{-2}, 10^{-3}, 10^{-4}, 10^{-5}

Another aspect may be extremely relevant with respect to the temperature dependence of switching data. While the agglomerates may grow considerably more at higher temperatures, the switching also becomes easier. This aspect will need another set of investigations on the interaction of domain walls with existing defects, which could not be given in the present work. An effect that is definitely due to an enhanced size of defects during fatigue is the decreased slope of the polarization hysteresis around E_c and the corresponding widening of the strain hysteresis minimum. This is easily explained in the framework of the model just given.

5.8.6 Rapid Fatigue due to Polarons

The temperature dependence of fatigue was also analyzed by Brennan et al. (1994). For the temperature range from 73 K to 423 K an exponential temperature dependence of the fatiguing rate was observed (20/80, 0.3 μm thick, Pt-contacts, same cycle scheme as Bornard et al. (2000), see p. 163). They arrive at an activation energy for the fatigue process of 0.043 eV. The rate of

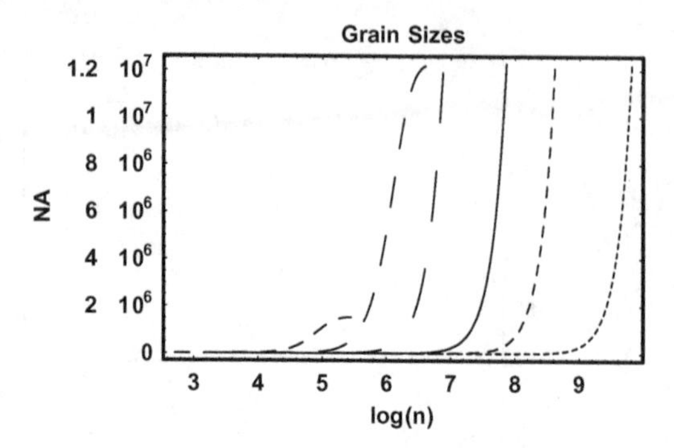

Fig. 5.13. Grain size dependence (equivalent to the inverse of initial agglomeration source density) of agglomeration according to (5.40) for (left to right): 0.5 µm, 1 µm, 2 µm, 5 µm, 10 µm, 30 µm

Fig. 5.14. Atomic force microscopic images of domain structures (**a**) before and (**b**) after 10^8 bipolar cycles at 2 E_c (image width 10 µm)

polarization reduction was:

$$R(T) = A \cdot \exp\left[\frac{-E_a}{k_B T}\right] . \tag{5.41}$$

Then the next step in their model was to assume that every ionized species would fix a certain amount of polarization. They arrive at

$$P_{ionized} = 1 - \left(1 - A \exp\left[\frac{-E_a}{k_B T}\right]\right)^{10^n} , \tag{5.42}$$

but they do not correlate this to the activation energy they found for the temperature dependence of the fatiguing rate (5.41). At their own account the description is not sufficient to describe fatigue. The fact of actually assuming a particular rate of thermally ionizing defects is not in reasonable agreement with the fact that all defects are frozen in at room temperature and essentially determined in their ionization state by high temperature sintering. Re-charging is only reasonable in account of the applied electric fields in conjunction with the remaining thermal energy at room temperature. An activation energy of 0.045 eV directly yields 90% ionization at room temperature and would only entail a cycle number dependence of the remaining 10% of switchable polarization.

A second model based on the electronic character of the fatigue process was recently presented by Tajiri and Nozawa (2001). The ionization of defects or host crystal ions is due to the fatigue process. The localized electronic carriers subsequently pin the domains in the sense above (Sect. 5.8.5). The authors developed a model that includes thermionic emission according to the Richardson-Dushman and field emission of electrons according to the Fowler-Nordheim equations (Hench and West, 1990). This model yields an extremely strong voltage dependence of the fatigue effect and basically no temperature dependence. They compare PZT to SBT (see Sect. 6.10) and explain the lower fatigue in SBT by the much larger mean free path of electrons in PZT. The electrons in the layered structures of SBT are confined to the perovskite layers, which yields a much reduced volume for acceleration. The impact ionization therefore is highly reduced. If a distribution of defect energy levels is assumed in the band-gap the appropriate loss of switchable polarization is found for cycle numbers observed in experiment.

5.9 Crystallite Orientation and Anisotropies

Crystallite Orientation

Two aspects of crystallite orientation are relevant in fatigue modelling. First it determines the projection of the electric field onto the crystal axes and thus the effective driving forces on domain switching. Secondly, considering the inverse effect, once defects have developed, they will at equal nominal size be more effectively blocking the domain wall motion, if they are located in not well oriented grains, because the domains experience lower driving forces than their more favorably oriented counterparts in other grains.

For the modelling of the fatigue process the orientation of the grains with respect to the external field direction has to be considered. As the data by Pan et al. (1992c) and our own results on the texture demonstrate, the development of defects interfering with the domain wall motion is not arbitrary in direction. Pan et al. (1992c) (for the discussion of their work see also pp. 51 and 73) reported that the fatigue in PZT is macroscopically highly anisotropic. The same is true for their X-ray data, where they were able to show that a higher alignment of c-axes was observed in the direction of the cycling electric field. The alignment was significantly higher than in the case of simple poling. They used a model of the interaction of a unit cell polarization with an oxygen vacancy to arrive at a high texture. This is almost the same picture as derived in Sect. 5.8, but their picture of single vacancies has the disadvantage to only consider isolated ions. In this case a motion of the vacancy to a next neighbor oxygen location is easily attained and thus the polarization may easily rotate along with it.

Field Directions in Single Crystals

The anisotropy of fatigue in single crystals and films was first discovered by Mansour and Vest (1992). A series of works by Takemura et al. (2000); Bornard et al. (2000) and Ozgul et al. (2001) followed recently. In rhombohedral relaxor $Pb(Zn_{1/3}Nb_{2/3})O_3$-$PbTiO_3$ (PZN-PT) single crystals the fatigue is highly anisotropic. Using the notations of quasi cubic (*qc*) unit cells, the fatigue along $[001]_{qc}$ ($P_r = 0.3\,C/m^2$) was retarded several orders of magnitude in cycle number (3.2 kV/mm, bipolar triangular, 10 Hz, 10^5 cycles $\hat{=}$2500 s at maximum field at one polarity) with respect to the equivalent case along $[111]_{qc}$ ($P_r = 0.45\,C/m^2$). In pure PZN crystals P_r dropped to 20% within 10^5 cycles along $[111]_{qc}$, in PZN-4.5% PT fatigue along $[111]_{qc}$ started already after 100 cycles and also dropped to 20% within 10^5 cycles, while no fatigue was observed up to 10^5 cycles along $[001]_{qc}$. Tetragonal PZT-12% PT, on the other hand, showed polarization fatigue down to 10% after less than 100 cycles (23°C) along $[001]_{qc}$, and similarly PZT-10%PT at -70°C. At this temperature PZT-10% PT also becomes tetragonal. For intermediate angles between $[111]_{qc}$ and $[001]_{qc}$ intermediate damage rates are observed.

Takemura et al. (2000) also reported recovery of the damage of about (60-70%) 80% of the initial remnant polarization after heat treatment for 16 h at (250°C) 450°C. The reduced fatigue rates at elevated cycling temperatures were another thermal effect. While at 23°C P_r in PZN-4.5%PT dropped to 25% of its initial value after 10^5 cycles practically no fatigue was observed at 100°C.

A further observation by Takemura et al. (2000) was the enhanced density of domain walls in PZN-4.5%PT after fatigue along $[111]_{qc}$, while fatigue along $[001]_{qc}$ reduced the number of domain walls. They explain the effect in the $[111]_{qc}$-oriented case by total pinning of some domains at defects. For subsequent switching to occur, new domain walls have to be generated at the same nucleation sites as previous domains, just the driving forces are slightly higher due to the already fixed domains. The very nature of the pinning centers is not discussed.

When the fatiguing voltage along $[001]_{qc}$ was raised to 4.9 kV/mm, significant fatigue occurred. This was ascribed to the field induced crystallographic phase transformation from rhombohedral to cubic, which occurs in PZN-

4.5%PT at around 2 kV/mm (along $[110]_{qc}$) and 4 kV/mm along $[001]_{qc}$. If driven up to 4.9 kV/mm, some parts of the electric field cycle occur in the tetragonal range.

The influence of temperature on fatigue was that PZN-4.5%PT degraded down to 25% of its initial P_r along $[111]_{qc}$, but showed practically no fatigue at 100°C. This is actually quite the opposite effect of the measurements by Brennan et al. (1994) (see above p. 5.8.6), who observed that fatigue rates in PZT thin films linearly increased from 100 K to 500 K.

A different influence of elevated temperature was anticipated, when it leads to a crystallographic phase change to tetragonal. Here again the higher fatigue rates in the tetragonal material are more relevant and were observed to rapidly stabilize the tetragonal structure also at room temperature.

Thin film $Pb[Yb_{1/2}Nb_{1/2}]O_3$-$PbTiO_3$ on $SrRuO_3$ oxide electrodes behaves similarly (fatigue at 50 kHz ($\hat{=}$25 kHz rectangular bipolar, with 8.6 µs at $\pm V_{max}$ and 20 µs unipolar cycle length), but the samples fatigue much later at 10^{10} cycles ($\hat{=}$86000 s under maximum field of one polarity, Bornard et al. (2000)). Other than the longer time until fatigue occurred, the results on film orientation are the same as those from the single crystals.

Implications

The crucial relevance of the domain system as part of the defect driving force can be seen from the fact, that the fatigue in PZT is so highly anisotropic as was shown in Sect. 2.4 and modelled in 5.8.5. As already pointed out by Sturman (1982), the local electric fields in a microstructure add up most effectively, just when the external applied electric field is opposing the field generated by the spontaneous polarization before switching, which is right beneath the coercive field. As an existing agglomerate fixes the neighboring domains to some extent, the fields for polarization switching may reach values much higher than for domains that are not yet fixed by some obstacle and thus much higher than the "local" coercive field of the single crystallite. This is not to be confused with the effect of the arbitrary orientation of the electric field with respect to the crystallite orientation.

Under bipolar cycling conditions, the fields due to the spontaneous polarization are fairly high. As all domains are allowed to move, at least at the beginning of the fatigue process, the locally existing space charges of whatever kind, electronic or ionic, will not move along with the moving domain walls. But they are needed to reduce the field due to the spontaneous polarization. Thus, under a.c. switching the field due to spontaneous polarization is not reduced and directly adds to the external field.

The 180° domain wall system around the defect, therefore, enhances the agglomeration process also via the enhanced defect drift in its vicinity upon field reversal, but the effect of the agglomerate upon fatigue is the even stronger modification of the mobility of the 90° domain walls when crossing it.

Despite the fact that all grains have switched from rhombohedral to tetragonal (see Sect. 3.3), the polarization and strain perpendicular to the fatiguing direction are almost not reduced after $2 \cdot 10^7$ cycles. This means that the shear stress exerted on the domain system by the agglomerates is high enough to change the crystal structure, but is still so highly oriented within the grains, that domains in the perpendicular direction move almost unaffectedly.

5.10 Relaxation Times

Slow bipolar cycling ($\approx 1.7\,\mathrm{mHz}$) at twice the coercive field yielded no fatigue effect in thin films during the first few cycles as already described in Sect. 1.13. Colla et al. (1997) used a second cycling condition at this low frequency, where the field was first raised to about the coercive field and then briefly increased to $2\,E_c$. Enormous fatigue was observed within a few cycles. Their assumption was that this is clearly an indication that the domain wall mobility is reduced in this case, because thin films only contain a substantial amount of domain walls close to the coercive field, while at high fields almost all grains are monodomain. Their conclusion is that fatigue in thin films can thus not be due to a blocking of domain wall motion, because they are always driven to the maximum field and just simply do not exist during long enough periods. They further support their model by measuring the dielectric permittivity at different bias fields. At high bias fields ($4 \cdot E_c$) the permittivity is raised while it is strongly lowered near E_c. As was discussed in Sect. 5.8.5 above, their data become very reasonable in the context of the agglomeration model developed here. The maximum ion flux occurs for fields just opposing the polarization right beneath E_c and induce the most fatal agglomeration possible.

If one considers the domain wall to move through a viscous medium, the motion is a priori blocked by minor obstacles, which are thermally surpassed at ease. The effective response of the medium is simply a slow down of the domain wall motion. As the agglomerates grow, the number of obstacles reduces linearly and their size increases. As the thermal activation drops exponentially the velocity of the domain wall also drops exponentially, even though the number of obstacles effectively decreases. Even if all obstacles can be surpassed thermally eventually, the rate of polarization reversal drops immensely. This is then reflected in the extreme rate dependence of polarization hysteresis and relaxation rate (Sect. 2.2 and 2.5).

5.11 Remaining Unclear Effects

The following effect has not been explained in literature, but noted to be peculiar, and is also not directly evident from the present work.

During the fatigue of thin films (PZT 47/53, 220 nm thick, Pt-electrodes, cycling voltages 15 and 30 kV/mm rectangular) Colla et al. (1995, 1998c) observed that for both cycling voltages the switchable polarization dropped to around 10% of its original value. But then after about $5 \cdot 10^6$ cycles at 30 kV/mm, the switchable polarization almost regained its full amplitude, while the fatigue at the lower voltage yielded no recovery. Along with this recovery an offset polarization was observed, which grew partially when the switching polarization regained its higher value. At a later fatigue stage both polarizations decreased again to very low values.

Complete recovery of the switching polarization was also observed for a sample already fatigued up to 10^9 cycles at 15 kV/mm, when subjected to the higher cycling field. Only along with the later fatigue also the dielectric permittivity significantly decreased at around 10^7 cycles for both cycling voltages and electric leakage conduction increased. This fact can actually be explained along the lines of agglomerates being large enough to block the domain wall motion at 15 kV/mm, but not at 30 kV/mm, but the re-increase of switchable polarization remains a peculiar effect.

Different films were used in their studies, each with different electrodes and one with a different lead stoichiometry (Pawlaczyk et al., 1995). Films (260-290 nm) which were prepared at nominally stoichiometric composition PZT 53/47 without lead excess, and evaporated gold electrodes ($P_{r,virgin} = 0.28\,C/m^2$) contained a thin layer (15-20 nm) of pyrochlore phase at the surface underneath the gold electrode. These films already fatigued after 10^4 cycles. At 10^6 cycles the switchable polarization degraded to less than 20% of its initial value. At about 10^8 cycles the remarkable "rejuvenation" of switchable polarization was observed. While the primary fatigue arose at around the same cycle numbers, the rejuvenation occurred earlier for films fatigued at higher voltages (8 V, 10 V, 12 V).

Films prepared with significant lead excess and Pt electrodes yielded a significantly different behavior. Samples for which the top electrodes were simply sputtered ($P_{r,virgin} = 0.44\,C/m^2$) exhibited the commonly encountered fatigue at around 10^6 cycles and no rejuvenation was obvious in the data. For films where the Pt electrodes were annealed after sputtering ($T_{anneal} = 450°C$ for 30 min., $P_{r,virgin} = 0.33\,C/m^2$) again early fatigue occurred. For lower cycling voltages (5 V) the fatigue was more drastic than for high cycling voltages (10 V down to 70% of the original polarization) again showing some recovery at 10^8 cycles. For intermediate voltages (7.5 V) the polarization only dropped to about 50% and showed some recovery at 10^7 cycles, but fatigued again beyond 10^8 cycles.

Some of the fatigued polarization could be recovered by applying d.c. fields. For very low cycle numbers (10^5) the gold sputtered films could be restored to $P_r = 0.15\,C/m^2$, but only after 1000 seconds, while samples fatigued to $2 \cdot 10^8$ recovered to $P_r = 0.18\,C/m^2$ within 2 cycles. Basically no recovery for the highly fatigued sample ($2 \cdot 10^9$ cycles) was observed, despite its higher switchable polarization. For the sample with annealed Pt electrodes higher restoring voltages above the cycling field yielded higher restored polarization, similar to our results from Sect. 2.3.

Another observation of the same authors was the increased coercive field for thinner films for a saturating maximum voltage (Pt-electrodes, Pb-excess, $E_{max} = 30\,\mathrm{kV/mm}$). This had previously been easily explained by a low dielectric constant passive layer beneath the electrodes, across which a significant potential drop is anticipated. Pawlaczyk et al. (1995) do not consider this approach valid and develop a model based on space charge layers. For low maximum fields, the coercive field was almost proportional to the applied maximum field and independent of film thickness. This is likely to be a dynamic effect, because the triangular measuring voltage pulses always were of equal length (0.5 ms).

Furthermore, an increase of the remnant polarization with maximum field was observed in the same films. The maximum of $P_r = 0.40\,\mathrm{C/m^2}$ was only obtained for fields as high as $70\,\mathrm{kV/mm}$.

Altogether the authors ascribe the fatigue effect to two types of space charge layers underneath the electrodes. The first one already develops during processing. The initial space charge layer develops due to band bending in the semiconducting PZT at the metal electrode. Extra space charges accumulate underneath the electrodes during fatigue, but can be removed by high d.c. fields. The changes in the dynamic properties of the films are ascribed to the altered domain nucleation behavior at the electrode. The additional space charge layers are made responsible for a lower nucleation probability of new domain walls. This in turn reduces the switchability of the film and yields the fatigue effect.

6 Recent Developments

The first version of this book was written towards the end of 2001. Since then more than 150 papers have appeared describing different aspects of fatigue in ferroelectrics, and some earlier publication only became widely accessible then. Most of the reports concern the processing of thin films and a brief statement is given on the durability of the newly produced material or device. Some works have given additional conceptually interesting results which are summarized in the following sections and reflected in the light of the other chapters of this book.

Even though this book is intended as a review, a few publications may still have escaped my attention. I apologize to these authors and admit that absolute completeness is impossible.

6.1 Material Modifications

6.1.1 Excess and Deficient Lead Oxide

A question not much discussed in the previous chapters is how much lead vacancies take part in the defect equilibria and fatigue. Friessnegg et al. (2000, 2001) found that the incorporation of lead vacancies into PZT is consistent with their positron annihilation data. They also observed an increase of defects upon annealing in reducing atmosphere, but could not uniquely assign a particular defect to this effect. A capture of positrons by the positively charged oxygen vacancies is much less likely than a capture by negatively charged defects due to the electrostatic repulsion between the positron and the defect. The authors nevertheless assume a certain attractive force between the positron and the oxygen vacancy. Even though not impossible this assignment is still very unlikely. A higher lead oxide content during preparation actually improves the fatigue resistance by an order of magnitude, but 10% excess lead have to be provided in order to render this treatment effective (Yang et al., 2000a).

The stability of the vacancy pair $\{V_O^{\bullet\bullet}\text{-}V_{Pb}''\}$ was investigated in a first principles theoretical study for a lead titanate lattice by Pöykkö and Chadi (2000). Lead vacancy relaxation occurred mainly in the *ac* plane. The binding energy for the defect pair was determined to be negative meaning that no pairs form. Only second nearest neighbors exhibited a slightly attractive force. If other defects shift the Fermi level into the upper quarter of the band gap, the binding energy becomes positive (exothermic pair formation). Thus, the liberation of PbO from the perovskite structure is a process involving predominantly surface evaporation of PbO and subsequent diffusion of individual vacancies into the bulk of the material. The influence of the lead vacancy on fatigue is small. The formation and stabilization of $\{V_O^{\bullet\bullet}\text{-}V_{Pb}''\}$ pairs near oxide electrodes thus seems to be a minor effect in contrast to other assumptions (Shannigrahi and Jang (2001), see Sect. 6.3.2). A shift of the Fermi level in these materials particularly due to defects, or due to the electrodes in very thin films may thsu strongly alter the probability of the formation of stable $\{V_O^{\bullet\bullet}\text{-}V_{Pb}''\}$ pairs.

6.1.2 Doping

Cerium doping improved the fatigue resistance of PZT by 1 to 2 orders of magnitude in cycle number (Majumder et al., 2000a). Some of the better fatigue resistance of oxide electroded films may stem from the diffusion of metal ions from these oxides into the ferroelectric itself. In $SrRuO_3$ electroded PZT the diffusion of Sr into the grain boundaries of the PZT was observed by Tsukada et al. (2000) along with the improved fatigue resistance due to oxide electrodes. The incorporation of silver was shown to not much modify the (admittedly poor) fatigue behavior of the investigated bulk PLZT (Zhang et al., 2001c).

6.1.3 Secondary Bulk Phases

A fairly interesting approach to overcome bulk fatigue was taken by the group of Longtu Li at Tsinghua University in Beijing. They added several different materials as secondary phases into PZT. In order to do so, they chose compositions which did not chemically react at the sintering temperatures for PZT. SBN ($SrBi_2Nb_2O_9$) (Zhang et al., 2001b,g), SBT Kim and Lee (2002), as well as ZrO_2 Zhang et al. (2001f) proved to improve the fatigue resistance of the materials considerably. The authors assign the improved fatigue resistance to the fact that oxygen diffusion is enhanced through grains of the other phase thus permitting for the local equilibration of defects, rather than an accumulation of these at some location in the microstructure. A second argument could be that subtle modifications of the grain boundary character occur due to some doping induced by the second phase as shown in Sect. 6.1.1. This latter point was not part of their discussion. For samples incorporating SBT the switchable polarization shrank to values approaching intrinsic values of SBT at concentration levels of incorporated SBT significant for fatigue resistance improvement.

6.1.4 Reduction of PZT, SBT, and BiT

Hydrogen is a well established reducing agent used in electronic chip production to protect the metallic and semiconducting materials from oxidation. When ferroelectric perovskites are introduced at this fabrication step for memory elements they evidently suffer the same reducing treatment particularly during metallization (Takatani et al., 1997). Two effects occur, first the formation of oxygen vacancies and possibly even the reduction of some cations to their metallic state (Scott et al., 2001), which is a well established effect for lead. This limits the metal electrode materials to the noble metals, which do not require so strongly reducing atmospheres during sputtering. Kanaya et al. (1999) found that this treatment directly induces imprint most likely due to the formation of oxygen vacancies at the electrode PZT interface with similar effects as those discussed at length in Chap. 5.8.2. The second effect is due to the incorporation of hydrogen into the ferroelectric, mostly as OH^--groups. These OH^--groups are more mobile than oxygen vacancies, but nevertheless do form clusters and may thus strongly interact with the mobile domain system (Scott et al., 2001). The argument by Kanaya et al. (1999) was further supported by measurements by Wu et al. (2000c), who found strong fatigue for films with top Pt electrodes applied in low oxygen partial pressure. In a similar sense the reincorporation of oxygen into the films by oxygen plasma treatment improved the fatigue of Pt electroded PZT films (Jang et al., 2000). Thin platinum oxide layers did not impede the considerable improvement the fatigue resistance of films with Pt electrodes produced under 70% oxygen atmosphere (Kim et al., 2000b).

In some cases too low an oxygen partial pressure will also deteriorate the performance of $La_{0.7}Sr_{0.3}MnO_3$ (LSMO) oxide electrodes even though the effect is much less pronounced than for $La_{0.5}Sr_{0.5}CoO_3$ (LSCO) (Wu et al., 2000b). IrO_x oxide electrodes yield a good fatigue resistance of the thin film devices even after Pt application under reducing atmosphere, but the imprint resistance becomes very poor. Similarly, $LaNiO_3$ was used as a reduction barrier for PZT during Pt application improving the fatigue resistance of the device significantly (Kim and Lee, 2002), but no imprint data were reported.

The layered perovskite $Bi_4Ti_3O_{12}$ (BiT) exhibits significant fatigue (at 10^7 cycles) as a ceramic (200 µm, 1 Hz - 2 kHz, 1000-4000 V), if annealed in reducing conditions for more than 90 min (400°C, N_2, 3% H_2). Thus high numbers of oxygen vacancies here also induce fatigue and thus shift the pinning to de-pinning balance in the layered perovskites more to the pinning side (Nagata et al., 2001), see Sect. 6.10 beneath.

A surprising fact was the observation that fatigue also occurs, if SBT ($Sr_{0.9}Bi_{2.1}Ta_2O_9$) films are annealed in pure oxygen. Jin et al. (2002) found that an oxygen deficiency develops in films annealed at temperatures above 800°C despite the presence of pure oxygen and significant fatigue was found (200 nm, ±5 V triangular, 10^{10} cycles, 1 MHz).

6.2 Grain Boundary and Grain Size

6.2.1 Grain Boundary

The grain boundary in polycrystalline ferroelectrics plays a crucial role in the global material and fatigue behavior, which has now been recognized by several authors (Klie and Browning, 2000; Lee et al., 2000c,a, 2001a; Verdier et al., 2003).

Klie and Browning (2000) investigated the grain boundaries in $SrTiO_3$ using EELS in a scanning transmission electron microscope. They found that an increased amount of oxygen vacancies is likely to be present in the grain boundary. At room temperature Ti^{3+} is only found in the grain boundary, while at elevated temperatures (724 K) an increased number of Ti^{3+} ions is also observed in the bulk in accordance with defect equilibrium (Vollmann et al., 1997). The high relevance of charges at the grain boundary was also evidenced from data for unipolar fatigue studies, where a rearrangement particularly of grain boundary charges was made responsible of the observed offsets after unipolar fatigue (Verdier et al., 2003).

Lee et al. (2000b, 2001b) performed one set of excellent measurements for determining the role of grain boundaries in thin film fatigue. They used a masking technique to produce a rectangular array of small PZT seeds on a substrate. From these seeds individual PZT grains grew to an almost rectangular array of grains each contacted by an individual platinum electrode. Single grain memory cells did not exhibit any fatigue for $2 \cdot 10^{11}$ cycles at 1 MHz, or $4.4 \cdot 10^7$ at 1 kHz, while cells containing one grain boundary degrade within 10^6 cycles. Electrodes covering two intersecting grain boundaries even induced fatigue within 10^4 cycles (Lee et al. (2000a); Lee and Joo (2002), 1 kHz/1 MHz, 200 nm, Pt electrodes, ± 10 V). The authors ascribe this behavior to the crucial role of grain boundaries as a defect location and transport path. This result is highly relevant with respect to the results on unipolar fatigue discussed in Sect. 2.6 and in (Verdier et al., 2003).

6.2.2 Grain Size

Another aspect not so much discussed so far is the influence of grain size on fatigue. Yan et al. (2001) presented a study for PZT films of two different thicknesses, one being 160 nm and one 400 nm each sintered at different temperatures (500°C to 650°C) in order to produce grain sizes of the films from 60 nm to 150 nm. Thinner films as well as larger grain sizes yielded earlier fatigue. The authors explain their data according to the ideas of Dawber and Scott (2000a) and the mechanisms developed in this book by the formation and agglomeration of $V_O^{\bullet\bullet}$-clusters predominantly underneath the electrodes. The assumption is that these charges near the electrodes modify the switching behavior particularly of those grains containing the agglomerate itself. Large grains thus represent a fairly large volume affected by the modified switching. For smaller grain sizes the total volume is considerably smaller, because only a thin layer at the electrode is then blocked.

6.3 Several Layers of Different Composition

6.3.1 Lead Excess Layers Near the Electrode

Yang et al. (2001, 2000a,b) deposited several layers of PZT for different degrees of excess lead in the precursors. Layers nominally lower in lead content near the electrodes exhibited a very much increased fatigue. It is unclear, whether only lead or lead oxide escaped the sample in large amounts during sintering, most likely the latter. This means that also sufficient oxygen vacancies are incorporated into the microstructure to yield the observed very early fatigue of the samples.

6.3.2 Multiple Ferroelectric/Antiferroelectric Layers, Buffer Layers

Kang et al. (2001) prepared PZT thin films containing a ferroelectric or antiferroelectric buffer layer to the platinum electrodes. The additional layer of the antiferroelectric lead zirconate resulted in strongly reduced remnant polarization, but an almost completely fatigue free behavior. Lead titanate on the other hand much improved the maximum achievable remnant polarization, but fatigue resistance was poor.

A similar result was obtained for a PZT buffer layer between Pt electrodes and the PLZT memory cell (300 nm, 1 MHz, 10^{11} cycles, 18 kV/mm, Shannigrahi and Jang (2001)). Two effects improved fatigue resistance, the preferential growth of <111>-oriented ferroelectric PLZT layers and the buffer layer function of PZT. Their argument for the improved fatigue resistance was that defect pairs of $\{V_O^{\bullet\bullet}\text{-}V_{Pb}''\}$ form, which are blocked in their motion by the PZT layer, but no proof was given for this assumption. Such an assumption contradicts the theoretical results by Pöykkö and Chadi (2000), Sect. 6.1.1.

In a similar sense extra layers of lead titanate on Pt electrodes significantly improved the fatigue resistance of PZT films (Ren et al., 2002). Fatigue free behavior was observed up to 10^{11} cycles (150 to 400 nm, 9 V). Bao et al. (2002b,a) chose a different material combination in order to improve the fatigue resistance of a memory maintaining the high polarization values of PZT, namely Pt/BLT/PZT/BLT/Pt. The fatigue resistance very much improved and simultaneously a good imprint resistance was observed.

Similar to the above mentioned buffer layers even a very thin MgO layer seems to be sufficient to avoid fatigue in memory cells, even when applied directly onto silicon (Kim and Basit, 2000).

6.4 New or Modified Modelling Approaches

6.4.1 Relation to aging

An update to their fatigue model was recently provided by Dawber et al. (2001) explaining the fatigue in thin film PZT by the formation of ordered oxide vacancy structures particularly underneath the electrodes. Other than specifying the location for the defect agglomeration, their model stayed basically identical to their previous approach (Dawber and Scott, 2001). The emphasis in this work was the temperature dependence of the fatigue effect. The increased fatigue for higher temperatures was reproduced in their model. For $BaTiO_3$ on the other hand, they used a completely different line of arguments. They applied the aging model by Arlt and Neumann (1988); Lohkämper et al. (1990); Robels and Arlt (1993) and Arlt et al. (1995) assuming that the fatigue procedure also induces a reordering of defect dipoles. The latter subsequently clamp domains. The inverse temperature dependence of the fatigue effect in $BaTiO_3$ with respect to PZT is explained this way due to a different temperature dependence of the dipole ordering. The data for $BaTiO_3$, on the other hand, concern a temperature range in which the spontaneous polarization is already strongly reduced due to the proximity of the Curie-temperature. Thus, arguments along the lines of the model discussed in Sect. 5.8.5 can also explain this effect. The ionic drift is smaller because the depolarizing electric field generated by the spontaneous polarization is considerably smaller when approaching the Curie point.

A strong temperature dependence of offset polarization was observed by Majumder et al. (2000b). Their thin films were fatigued at room temperature (PZT, 700 nm, 1 MHz) and subsequently measured at higher temperature (RT up to 125°C). A strong offset was observed indicating enhanced aging of the fatigued samples into a biased state. In their system the offsets seem to be even stronger than in our data (Sect. 2.2). Aging is thus a very relevant mechanism following the initial fatigue.

6.4.2 Ising and Preisach Approach

Lo and Chen (2001a,b) modelled the influence of oxygen vacancies on the single unit cell level like Park and Chadi (1998) using an Ising model approach. The frequency and amplitude response of ferroelectrics are reproduced. The simulated formation of a space charge layer modifies the local potentials such that the fatigue induced changes of polarization hysteresis and increased coercive field are found.

A similar mathematical treatment of the fatigue effect was chosen by Gondro et al. (2000) allowing for more than two polarization states in a Preisach approach. In a sense the model is identical to previous assumptions about a certain number of fixed or pinned volume elements. Their individual coercive fields change due to the fatigue process and the volume average is used to calculate global effects.

6.4.3 Phase Transition Picture

The agglomeration of large quantities of oxygen vacancies can lead to structural phase transitions in the perovskites. In $SrTiO_3$, e.g., a percolative phase transition was identified at a critical concentration of around 7% vacancies. Scott (2001) used this fact to interpret the agglomeration of defects and the changes of material properties in PZT due to fatigue (Scott and Dawber, 2000a). To what extent such a picture can be considered at the much lower defect concentrations in PZT is not clear. Considering the annealing studies of Sect. 2.3.2 where a fairly high recovery of switchable polarization is obtained due to annealing the fatigued phase can be considered the thermodynamically stable phase at room temperature. The unfatigued phase is metastable but kinetically frozen in all materials commonly encountered. At higher temperature, where no spontaneous polarization exists, the thermodynamically stable phase at this temperature is reached within reasonable times during the annealing.

When X-ray data are considered, each domain state can be identified as a separate of several thermodynamic phases. The fact that a certain domain orientation becomes a preferred one due to the fatigue process can be considered a first order phase transition between two phases. Thompson et al. (2001) identified the pinning of one preferred domain orientation (Sect. 5.8.5) in their X-ray data. The two states were termed easy and difficult domain states, out of which only the easy survives the fatigue procedure. In this sense one could also consider the pinning of domains a phase transition process.

6.4.4 Supplemental Arguments to Previous Chapters

A comprehensive summary of works from the Lausanne group was provided by (Tagantsev et al., 2001). This review in a very broad sense states several of the arguments used in this book, but arrives at a different conclusion. A discrimination between possible fatigue mechanisms is provided. The first concerns electrode delamination as described in Sect. 3.8. This mechanism equally affects the maximum obtainable polarization and the coercive fields needed for switching. A second set of mechanisms involves the reduction of the fields locally present and thus an apparent higher coercive field. The third invokes more difficult domain switching itself, which has been the major emphasis of this book. The major experimental findings beyond the facts already discussed concern the leakage current (see Sect. 6.5) and the influence of frequency on fatigue. The leakage current in thin films is reduced, which is not the case in the bulk material used here. It is unclear whether this is due to the fact, that a different composition of PZT was used in their study, or whether this is actually the influence of the electrodes which comes into play for the thin film. All mentioned fatigue mechanisms can be made responsible for the commonly observed slanting of the ferroelectric hysteresis loop. The discussion in the paper then focusses on the distinction between a scenario invoking the redistribution of defects within the material or a charge injection picture. A mechanism which is equally well probable but not discussed in their work is the simple ionization of existing defects in the bulk. This similarly entails an increased number of defects capable of pinning domains. As far as I am concerned, I think all three mechanisms are relevant. The major argument in favor of charge injection necessitates a thin dielectric layer across which an over-proportionally high electric field develops, once the polarization in the adjacent ferroelectric has switched. Once the material has fully switched, the local fields in this dielectric layer become small again. The field argument is thus the same as in Sect. 5.8.5, but the localization is very different. While in the assumptions in this book the unscreened fields develop within the ferroelectric, they are here seen as fields mostly relevant in the dielectric layer. These high fields then induce the electronic charge injection responsible for the fatigue effect. This picture is further supported by the size effect of measured coercive fields. The arguments in the paper by Tagantsev finalize in the seed inhibition mechanism for domain switching. This mechanism considers the growth of domain wall seeds to be suppressed near the electrodes due to injected charges. I think with the present state of measurement both mechanisms may equally well be valid in thin films. A similar model was then further extended by Grossmann et al. (2002a,b). They also argue that a thin dielectric layer exists at the electrodes, which entails charge rearrangement and defect dipole reorientation in the bulk and thus the observed bias and reduced switchability of domains. In bulk ferroelectrics as discussed in most parts of this book, the seed inhibition can not be the dominant mechanism, because fatigue was observed throughout the entire microstructure (Sect. 3.1). As stated by Tagantsev et al. (2001), an increased density of domain walls is a necessary consequence of domain pinning mechanisms within the bulk. As was shown in Sect. 3.4 the density of domains is highly increased due to fatigue. Thus a bulk fatigue mechanism definitely prevails in bulk PZT.

A very different picture of the electron injection mechanism has recently been developed by Dawber et al. (2002) and Dawber and Scott (2002) evoking screening effects within the metallic electrodes and subsequent electron tunneling. The remarkable outcome of this model is that there is no necessity of a "dead layer" (layer of non ferroelectric low dielectric material) near the electrode which is the assumption in all of Tagantsev's works. For Dawber et al. all charge injection occurs simply due to the fact, that the externally applied electric fields can not fully develop at the surface of the ferroelectric due to a finite screening length within the metal electrode itself ($\lesssim 1\,\text{Å}$). For a sufficiently small layer at a metal surface the electron density within the metal is not sufficient to fully screen the electric field. Thus, some potential drop also occurs in the metal. According to their numerical estimate this effect is sufficient to lift the necessity of a dead layer. Unless processing-related, the formation of a dead layer near the electrode seems unlikely according to their argument.

A supplemental study of the fatigue model by Dawber and Scott (2000a), Sect. 5.8.4, was recently provided by Wang et al. (2002). They used the identical model, but introduced a few arithmetic corrections. The general trend in the material behavior remains unchanged, but the match between the experimental and theoretical curves is not as good any more.

The influence of the formation of charged layers within a bulk ferroelectric on the macroscopic material behavior was recently investigated by Bratkovsky and Levanyuk (2000b). This fundamental theoretical work is based on Maxwell's field equations and a Landau-Devonshire approach for material description, particularly with respect to phase transitions. In their model the influence of free charges on the switching behavior is developed from a free energy in terms of polarization (no strain coupling). An intrinsic polarization associated with the free charges develops. The modified material response is then treated for different constitutive relations. In a linear dielectric the intrinsic polarization introduced by the free charges is independent of temperature and the associated changes in the dielectric constant. The dielectric constant on the other hand diverges at the phase transition. For a ferroelectric the transition temperature (Curie-point) drops due to the additional charges by 1 to 100 K using standard material parameters. A built-in field develops, which is distinctly different from an external field. According to their calculations, an external field would smear out the phase transition temperature. Like in many of the fatigue pictures the polarization splits into two fractions, one non-switchable and the other one remaining switchable. The latter becomes the order parameter determining the phase transition temperature. Its modified temperature dependence strongly changes the material parameters also at room temperature. The additional free charges in the material surprisingly do not alter the coercive field. The entire treatment considers the thermodynamic parameters, particularly the thermodynamic coercive field. This value is much larger than the experimental coercive field, because experimental switching takes place as a nucleation and growth process and not via homogenous switching, which is determined by the thermodynamic coercive field. The surprising result of their work is, that overall the coercive field should drop due to fatigue and not increase like mostly observed. This is due to the fact, that the phase transition temperature is so highly reduced and, accordingly, in the Landau-Devonshire approach the coercive field drops. For very thin films the nucleation and growth mechanism becomes less and less likely until for very thin films ($10^4 - 10^6$ Å) it is impossible and the material behavior approaches the thermodynamic switching behavior and the coercive fields calculated here. Most of their development is formulated for a material exhibiting a second order phase transition, but the authors claim is that it is equally valid for first order phase transitions. According to their analysis, the explanation of the enhanced coercive fields due to space charges within the material or in the electrode proximity as proposed by Tagantsev et al. (2001) is not plausible. Without specification of the particular switching mechanism and its modifications due to fatigue, they consider such statements not meaningful.

6.4.5 Curved Domain Walls

The arguments used in Sect. 5.8.5 for the development and mobility of needle domains were recently matched by a set of beautiful TEM studies on $BaTiO_3$ and $KNbO_3$ by Krishnan et al. (2000b,a, 2002). They observed that charged domain walls like the needle tips in Fig. 5.4 moved at a much higher velocity under the application of external fields than compensated domain walls. Particularly, geometrically small domain patterns moved first followed by larger domains. Charge-neutral domain configurations only shifted after application of considerably higher fields. As a consequence the authors argue that the number of stable charge-neutral configurations is enhanced after fatigue. Their stability determines the reduced switchability of the entire domain system. This is the same argument as used throughout Sect. 5.8.

A set of experiments further supporting a bulk scenario of fatigue was provided by Chou et al. (2001). They observed in $(Pb_{1-x}Sr_x)TiO_3$ that the initially few domains multiply due to bipolar fields applied to their samples. 90° domain walls dissociate into an increasing amount of interlocking zig-zagged domains finally leading to a severe reduction of switchable polarization. Even dislocations were found.

6.4.6 Other Approaches

Ricinschi et al. (2000) developed a simple model relating the hysteresis loop parameters to the fatigue state based on a Landau-Devonshire free energy formalism. They found that a simple reduction of switchable polarization is not appropriate to describe the observed hysteresis loops after fatigue and that a second fraction of grains has to exist which experiences a modified local field. This modification yields an improved fit of the experimentally observed data, but is not fully sufficient in the authors own consideration.

6.5 Leakage Current

6.5.1 Ionic Currents

Majumder et al. (2001) performed an extensive study on leakage current in PZT (53/47) doped with different amounts of iron. All leakage current contributions were thermally activated. A reduced leakage current was found for iron doping up to 4 at.% along with an improved fatigue resistance. The authors associate the improved fatigue resistance with the formation of Fe^{3+}-$V_O^{\bullet\bullet}$ vacancy pairs, which immobilize the oxygen vacancies. Thus redistribution of the vacancies due to the fatigue process becomes less and less possible for increased iron doping.

6.5.2 Electronic States

Scott and Dawber (2002) argue from infrared absorption data ($4000\,cm^{-1}$), that the electrons liberated from oxygen vacancies are highly localized (deep trap or small polarons due to Anderson localization (Tsuda et al., 2000)). They further support their and the model of Sect. 5.8 for oxygen vacancy agglomeration. For thin films they assume the agglomeration to predominantly take place underneath the electrodes. They use previous data by Shimada et al. (1996) who found oxygen vacancy clustering from current voltage studies at different temperatures in $Ba_{0.7}Sr_{0.3}TiO_3$. Nozawa et al. (2001) on the other hand discuss that the pinning defects are not necessarily oxygen vacancies, but also consider electronic trap states. Thermionic emission of free charges from these localized levels is assumed as the underlying mechanism. A wide distribution of electronic trap levels in the band-gap is needed for the modelling in order to reproduce experimental data. Some of the distributions widths even exceed the band gap, which seems somewhat unreasonable.

A leakage current study considering electronic band gap states as the major itinerary for charge currents was provided by Nagaraj et al. (2001), see also Sect. 6.5.3. For Pt electrodes the leakage current linearly decreases on a log scale, while hardly any changes in leakage current occur for oxide electrodes (LSCO). The electron emission process from the Pt electrodes at low temperature is assigned to Poole-Frenkel emission and at higher temperature to Schottky emission. This result was obtained by comparing the dielectric constants for high and low frequencies. At room temperature the leakage current measurements proved difficult due to polarization relaxation currents. While for short times an asymmetry was observed, the steady state currents showed no polarity dependence. The current relaxation observed with Pt electrodes was assigned to charge entrapment. As the power law relaxation of the steady state currents was independent on temperature, the density of traps in the band gap was assigned the relevant material parameter dominating the charge capture relaxation process. Oxygen vacancies as well as $Ti^{4+} \rightarrow Ti^{3+}$ were considered possible trapping sites and the similarity to fatigue effects was pointed out.

A similar argument invoking the better fatigue resistance of PZT thin films with oxide electrodes is the modified distribution of electronic states in the band gap (Stolichnov et al., 2001a). The formation of lead ruthenate ($PbRuO_3$) at the interface between RuO_x and PZT during sintering allows for charge relaxation in this area due to the different set of trap levels in the band gap. The basic argument for this scenario is derived from the thickness dependence of the coercive field in materials with metallic electrodes.

There was no microscopic evidence given for the existence of $PbRuO_3$ and the relevance of such a layer is actually questionable.

6.5.3 Breakdown

An interesting study relating the dielectric breakdown to the fatigue state was provided by the Lausanne group (Stolichnov et al., 2001b). The leakage charge to breakdown proved to be a characteristic material parameter independent of applied voltage but not the leakage current. For $SrRuO_3$-electrodes the charge to breakdown was 10 times higher than for Pt-electrodes. 10^6 bipolar fatigue cycles reduced the charge to breakdown value by another factor of 4 (6 V, 25 kHz, on PLZT 110/1,5/45/55 thin films of 300 nm). They observed that fatigue yielded much lower leakage currents (actually an observation contrary to our own), which would prolong device lifetime, but also much lower charge to breakdown which is counterproductive. Hole injection is considered the relevant parameter which via impact ionization generates charged defect states in the bandgap subsequently percolating to become the conduction path (Stolichnov and Tagantsev, 1998, 1999). Simple valence band conduction is excluded according to their arguments, because this would yield breakdown independent on polarity, which is not the case. The breakdown is assumed to occur in the bulk of the ceramic and not in the grain boundary, because of the low scatter in the data. The $SrRuO_3$ electroded material is so much more breakdown resistant, because strontium is incorporated into the grain boundary and provides current paths for the observed high currents. This shields the fields from the breakdown sensitive bulk. The fatigue generates a non-uniform distribution of current between non-fatigued and fatigued regions yielding the earlier breakdown. Even though the discussion is very speculative, it appears reasonable, but many more data have to be collected to make it a viable scenario.

6.6 Mechanical Effects

6.6.1 Mechanical Fatigue

Hauke et al. (2000, 2001) examined the fatigue behavior of thin films deposited onto steel under transverse mechanical loading. Their cycling procedure imposed transverse strain onto the PZT layer by bending the metallic substrate. Mechanical cycling at 1 Hz frequency and imposed strain values of $\Delta s_{11} = 0.15\%$ (and $\Delta s_{22} = 0$) yielded strong material fatigue after 10^5 cycles, while bipolar electrical cycling at 50 Hz, 40 kV/mm $\approx 5\ E_c$ deteriorated the identical films after 10^6 cycles. In the electrical case the material was transversely fully clamped (in the mechanical case only in 2-direction). Thus only 90° switching was permitted in those grains, for which no or little change in strain values would occur in the plane of the steel substrate. Many grains under these conditions only permit purely 180° switching. After electrical fatigue the strain and polarization hysteresis loops both stayed symmetric despite the two different metal electrodes and the commonly observed offsets as discussed in Sect. 2.3. The authors assigned all loss of switchable polarization to the formation of microcracks, but did not explicitly observe any. Their explanation seems right, but the purely macroscopic observations cannot fully prove this mechanism. A simple point defect migration scenario is also possible. In the electrical case the same mechanisms as discussed in Sect. 5.8 with the offset only weakly developed may be present here. An enormous increase in coercive field was observed after electrical cycling, much higher than in most other thin film studies and particularly more than for macroscopic bulk samples. In the mechanical case no hysteresis loops were shown after the fatigue treatment. The microscopic mechanism can still be identical to the electrical one discussed in the same Sect. 5.8, in the case of open circuit cycling, because then the depolarizing field fully develops macroscopically. A definite assignment is not possible, because the electrical boundary conditions are not specified for the mechanical fatigue experiment.

6.6.2 Microcracking

Liu et al. (2001) identified similar patterns as those described in Sect. 3.1 as etch grooves. These structures were found in thin film PZT fatigued to 10^{11} bipolar cycles using liquid indium electrodes, which were easily removed after the fatigue treatment, because the liquid indium is not wetting the PZT and thus drips off the surface of the ferroelectric by itself. The structures observed in the atomic force microscope were interpreted as microcrack trees generated from the strain mismatch at 90° domain walls. Similar structures were also found in one of our own studies by atomic force microscopy (Lupascu and Rabe, 2002), but interpreted as agglomerated point defects.

The dielectric breakdown of ferroelectric films was assigned to be dominantly due to mechanical effects (Sun and Chen, 2000). In all cases, where microcracks formed in thin ferroelectic films, an enhanced susceptibility to breakdown was observed in these films. The recommendation by the authors was to use materials with small electrostrictive coefficients for memory cells.

A beautiful TEM study revealed the close correlation between microcrack formation and the stress field near triple points and along domain walls (Tan et al., 2001). In PZT the strong fields at the crack tip of already existing cracks induce complete switching of the neighboring grains. These in turn exert forces on the other crack face wedging open the crack tip at those neighboring grain locations, where the field is not changing the domain structure as strongly. Similar to the results in Sect. 3.7.2 the same authors observe the formation of a glassy phase at the grain boundaries and triple points, which weakens them electrically and mechanically. Cracks predominantly propagate along grain boundaries.

6.7 Anisotropy of Fatigue

Chu and Fox (2001) were able to extend the results by Bornard et al. (2000), p. 162, to the PLZT system. Similar to the relaxor system PZN-PT, PLZT exhibits improved fatigue properties if cycled along the [001] direction. PLZT films containing around 10% of <001> oriented grains best improved the fatigue resistance. While purely <111> oriented films degraded by 90% within 10^9 cycles, the switchable polarization only reduced by 20% for additional 10% <001> oriented grains. Epitaxial thin films of $Pb[Yb_{1/2}Nb_{1/2}]O_3$-$PbTiO_3$ with $SrRuO_3$ oxide electrodes grown on <001> oriented $LaAlO_3$ or <111> $SrTiO_3$ substrates also show a distinct fatigue anisotropy (Bornand and Trolier-McKinstry, 2000). While the <001> oriented films are essentially fatigue free up to 10^{11} bipolar cycles <111> oriented films fatigue rapidly. The unbalanced 110°/70° domain wall system in the <111> oriented system was suggested to exhibit stronger interaction with charged defects and thus a higher susceptiblity to fatigue. The same orientational behavior can also be found for $Pb_{0.99}[(Zr_{0.6}Sn_{0.4})_{0.85}Ti_{0.15}]_{0.98}Nb_{0.02}O_3$ grown on textured $Pb(Zr_{1-x}Ti_x)O_3$ seed layers (Yoon et al., 2002). <001> textured films are fatigue free, while <111> films significantly fatigue within 10^9 cycles (300 nm, ±10 V, 100 kHz).

In another set of orientations Kim et al. (2001b) found that the in-plane tetragonal polarization direction [100], which is initially perpendicular to the cycling field direction shows even lower fatigue resistance than the [111] orientation

An anisotropy of a different fashion was observed by Kimura et al. (2002) using synchrotron XRD on PZT. During fatigue cycling the texturing of the material increased along the cycling direction along with a a slight increase in lattice spacing.

6.8 Time and Relaxation Effects

Lee et al. (2001c) performed a low frequency study of the response of PZT (53/47) films (300 nm) with platinum bottom and gold top electrodes before and after fatigue (10^9 cycles, 50 kHz). At low frequencies (<1 Hz) the dielectric constant as well as the dielectric losses increased considerably due to fatigue, while for higher frequencies a reduction in dielectric constant was anticipated with approximately equal loss factors before and after fatigue. The imaginary dielectric constant expressing the losses in the material showed a distinct frequency peak around 7 Hz developing for rising temperature (up to 300°C). This peak was associated with mobile charge carriers. The rotational motion of defect dipoles as well as itinerant charge relaxation were discussed as the possible local mechanisms, with an extended discussion of the relaxation reorientation of the vacancy-pair $\{V''_{Pb}\text{-}V^{\bullet\bullet}_{O}\}^{\times}$. An exquisite result of their study was the determination of two distinct activation energies of 1.61 eV before and 0.96 eV after the fatigue process.

An obvious indication of the modification of the domain dynamics also in thin films was given by Kang et al. (2003). They studied the relaxation behavior of PZT after fatigue. A change of the polarization state with time was observed for fatigued samples entailing the retention loss of the information written into the ferroelectric. A power law time dependence for the loss in discernable positive and negative polarization states was observed representative of a distribution in time constants. Two explanations are possible to explain this effect. First, there is a back-driving force on the domain system induced by fatigue not existing in the initial material. The second one is an aging effect that occurs after fatigue due to an enhanced availability of charged defects. Similar to the assumptions in Sect. 5.8.5 the driving force for this time dependence is assumed to be the unscreened depolarizing field.

The fact that fatigue times rather than frequencies are the relevant parameter for the fatigue process was demonstrated by a simple observation of a master curve for the frequency dependence of fatigue in PZT . For their Ceria doped PZT films Majumder et al. (2000a) found that all polarization loss can be reduced to a common master curve of polarization vs. N/f^2 with N the number of cycles and f the frequency. The loss of switchable polarization follows a stretched exponential behavior.

The dynamics of oxygen vacancies was made responsible for the same effect by Zhang et al. (2001e). They found a strong frequency dependence of fatigue in PLZT 2/70/30 and also measured the activation energy for the high temperature conductivity to be 0.95 eV (540°C through 570°C), which is very close to the value previously reported by Warren et al. (1996c). Like in the work by Majumder et al. (2000a) the material is not fatiguing at higher frequencies (>5 kHz, $N_{max} = 5 \cdot 10^8$). On the other hand the data may be highly misleading, because they were measured on bulk ceramics. No statement was given on the thermal behavior. In our own bulk samples a frequency of 50 Hz heats the samples to around 70-90°C. Possibly all high frequency data are measured on hot samples which are not ferroelectric any more. No statement on sample dimension and thermal conditions is given. At the same time these authors found a substantial increase of the dielectric constant in the paraelectric high temperature phase after fatigue Zhang et al. (2001d). My interpretation of these data is that the induced defects are still present in the high temperature, as shown by the annealing study in Fig. 2.9. Around these defects some needle domains are still present stabilized in the paraelectric phase by the defect agglomerates. The induced needles are highly mobile and can easily grow or shrink. Large polarization changes are induced much larger than by the paraelectric phase itself. This yields the increased dielectric constant of the fatigued phase.

The inverse effect was observed for thin film PT/PZT/Pt films (Sun et al., 2002), where a reduction of the dielectric constant and the Raleigh constant

(Damjanovic, 1998) was observed after fatigue. In this case an increased pinning of the domains also on a local scale has to be present. The authors attribute the reduced capacitance values to domain pinning at the electrode interfaces.

6.9 Electrode Geometry

Torii et al. (2001) investigated the influence of electrode edge effects on the fatigue behavior of PZT by applying comb-shaped Pt top electrodes of different electrode widths (0.7 μm to 5 μm) to thin film PZT. Using atomic force microscopy (AFM) they found that up and down domains are pinned at different stages of the fatigue procedure. In a late stage of fatigue (10^9 cycles) both types equally well occur, while for intermediate cycle numbers (10^7) first up-domains (poling positive) and then down domains (10^8) are pinned which corresponds to our observation of the macroscopic strain hysteresis asymmetry. This asymmetry was also most pronounced at intermediate cycle numbers (10^7) and then decreased again at 10^8 cycles (Sect. 2.3.1). For Torii et al. the pinning of domains underneath the electrodes is significantly stronger than in between the electrodes. Whether this is an effect of the blocking function of the metallic electrode for oxygen diffusion or whether it concerns only geometrical effects is not separated. Rejuvenation occurs in an intermediate stage of fatigue (see Sect. 5.11). Their interpretation is that a ferroelectrically non-active region exists, which screens some of the external field due to an enhanced drop of electric potential across this layer. The rejuvenation is then explained by the growth of conducting paths through this near electrode non-ferroelectric region. The early fatigue of material underneath their electrodes is explained by locally enhanced tensile mechanical stresses.

6.10 Layered Perovskite Ferroelectrics

A major breakthrough for memory applications was achieved when the almost fatigue free behavior of the layered perovskite ferroelectrics was discovered Scott (2000). Nevertheless, some fatigue and particularly aging occurs in these materials, if the boundary conditions and the processing are not appropriate.

6.10.1 Strontium Bismuth Tantalate

While in SBT ($SrBi_2Ta_2O_9$) no fatigue seems to be present at room temperature somewhat significant aging was observed in this layered perovskite (Kim et al., 2001a). Particularly at elevated temperatures (200°C) the switchable polarization considerably ages within a month. Accordingly, a.c. switching instead of fatiguing the polarization de-ages the material at these elevated temperatures recovering the switchable polarization by a factor of three from the aged state. Even though very fatigue resistant, the layered perovskites are prone to detrimental aging, if stored at elevated temperatures, which e.g. may limit the use of these materials in applications like cars etc., where high temperatures may be present during extended times of the devices lifespan. Significant room temperature aging of SBT films was observed for sputtered films and metal organic vapor deposited films, the latter showing a much stronger effect (Chen and Tuan, 2000). The aged films necessitated a similar number of bipolar cycles to be de-aged (500 nm, ± 5 V, 1 MHz) as other materials exhibit for fatigue. The low voltage is actually not able to provide sufficient de-pinning (see beneath).

A different explanation for the fatigue free behavior of SBT was given by Ding et al. (2001c). The particular domain structure in this compound permits 5 different domain or anti-phase boundaries. The latter are considered the crucial mobility advantage of SBT with respect to PZT, because 90° domain walls nucleate at these boundaries and exhibit a very high mobility. While in PZT charged domain walls strongly interact with charged defects, anti-phase boundaries do much less. The anti-phase boundaries thus serve as a multiple nucleation site for 90° domain walls. In this TEM study four out of the five possible domain/anti-phase boundaries were directly observed. Pure $Bi_4Ta_3O_{12}$ (BTO) does not exhibit anti-phase boundaries (Ding et al., 2001a). Slightly lanthanum doped BTO on the other hand does contain anti-phase boundaries (Ding et al., 2001b), which explains the fatigue free behavior also in this material (Park et al., 1999).

In an earlier study Dimos et al. (1996) described fatigue of SBT under broad band illumination. Their comparison of SBT (Pt) and PZT on LSCO electrodes showed that both systems exhibit strong fatigue, if illuminated (400nm, $\pm$ 4 V or $\pm$ 10 V). Particularly lower cycling voltages yielded very strong fatigue rates for SBT. Under high d.c. bias or cycling in the dark this fatigue state was easily recovered. The authors concluded that fatigue is essentially a competition between pinning and unpinning. The particular role of oxygen vacancies was to provide ionization sites located underneath the electrode, which were brought to this location during the initial fatigue in metal electroded PZT. Samples containing these vacancies were prone to higher fatigue rates after rejuvenation, because the latter were recharged due to illumination, but not reordered. Their location underneath the electrodes did not change by illumination. The electronic nature of pinning was emphasized by the common observation of this effect in SBT as well as PZT, keeping in mind that appropriate ionizable defects have to exist, e.g. oxygen vacancies or dopants. A similar argument was arrived at by Palanduz and Smyth (2000). They argued from thermopower and high temperature conductivity data, that the strong fatigue resistance of SBT stems from the large bandgap and the resulting low mobility of defects due to their deep levels within this bandgap.

A further study supporting the pinning unpinning picture was provided by Wu et al. (2000a). They applied rectangular voltage pulses of different amplitude and width to SBT. For long pulse widths and low applied fields more than 10% loss of non-volatile polarization (assuming this is the switchable polarization) occurs within 10^{11} cycles for 440 nm thick films.

A very large switchable polarization was observed for the layered perovskite $Bi_{4-x}Nd_xTi_3O_{12}$ (BNdT) along its c-axis (280 nm, $x = 0.85$, Chon et al. (2002)). This compound also shows no fatigue but the measurement procedure may be somewhat misleading, because higher fields were applied during measurement (±5 V) than during cyclic loading (±3 V). This may yield de-pinning at the higher driving force, even though some fatigue may occur for the identical read and cycling voltage, which would be the case in a memory cell.

6.10.2 Bismuth Titanate

Compounds of the bismuth titanate family of perovskite ferroelectrics have recently been considered as one of the materials replacing PZT in ferroelectric memories (Scott and Dawber, 2000b). While all SBT's seem to be resistant against fatigue even in ferroelectric memory cells containing platinum electrodes, pure bismuth titanate ($Bi_4Ti_3O_{12}$, BTO) shows similar fatigue as PZT. In lanthanum substituted BTO namely $Bi_{3.25}La_{0.75}Ti_3O_{12}$ (BLT) fatigue free behavior is observed and assigned to an increased stability of the oxide octahedron layers which was supported by XPS measurements (Noh et al., 2000). The particular role of lanthanum in stabilizing the oxide layers was noted. Similarly fatigue free behavior for BLT was found by Hwang et al. (2001); Tanaka et al. (2001).

A comparison of BLT and SBT concerning the fatigue properties was provided by Wu et al. (2001). The fact that no fatigue is observed in SBT had previously been assigned to a high degree of de-pinning of domain walls sufficient to release the domain walls again almost instantaneously after the initial pinning (Al-Shareef et al., 1996c; Dimos et al., 1996). According to Wu et al. (2000a) this also occurs for BLT for sufficiently high fields. Lower cycling voltages are less effective in de-pinning the domain walls and significant fatigue occurs. The voltage dependence in SBT is small and thus the trapping centers for the assumed electronic charge carriers are low traps, while in BLT the effect is stronger and deeper traps are assumed. The double charge of the oxygen vacancy with respect to electronic carriers along with the high volatility of bismuth oxide leads the authors to the assumption that the trapping centers are also oxygen vacancies in this material.

6.11 Uniaxial Ferroelectrics

Li et al. (2001) demonstrated that the well known uniaxial ferroelectric lead germanate ($Pb_5Ge_3O_{11}$) is also a possible candidate for memory cells in thin film deposition technique (thickness 300 nm). Fatigue at 80 kHz and $\approx 2.5\ E_c$ bipolar square waves did not induce any fatigue up to 10^9 cycles. The authors assign the absence of fatigue to the fact, that no 90° domain walls and very little electromechanical coupling exist. Point defect drift similar to Chap. 5.8 is also much less likely than in PZT because the spontaneous polarization is only around 10% of that of PZT. Lead germanate seems fatigue safe, but the remanent polarization is only around 10% of that in PZT. It has therefore so far not been seriously considered in memory device manufacturing.

7 Summary

The present document provides a summary of fatigue effects in PZT found in measurements by the author and interpreted in the framework of a large body of literature existing in this and neighboring subjects. Surprisingly, the knowledge on bipolar fatigue of bulk material is by far not as advanced as in thin films. The intention of this work is to arrive at a thorough understanding of the underlying physical mechanisms of fatigue and to outline how potential models have to be structured to be meaningful with respect to the microscopic mechanisms. A simple model for defect agglomeration is given based on the experimental findings of the agglomeration mechanism. A comfortable agreement with the experimental results, several of which were unclear so far was obtained without using any fitting parameters.

Fatigue is so complex in PZT, because it *does* correlate with all parameters and scales that influence ferroelectric behavior in general as was already hypothesized in the preface. The smallest scale found to be significant in fatigue is the mobile point defect. Already a single point defect changes the ferroelectricity of single unit cells and its surroundings. Isolated dopants as well as oxygen vacancies are found not to significantly alter the material behavior on a global scale, though. Only after agglomeration of some kind changes of the macroscopic behavior are anticipated. Agglomeration to small entities like defect dipoles arises particularly for acceptor doped PZT. This well known fact leads to aging in all of its forms. A mechanism arising due to bipolar fatigue, identified and experimentally supported in this work, is the agglomeration on the scale of the microstructure. It is this agglomeration that constitutes the major finding of this work.

Agglomeration, the Most Stable Defect Type:

Depending on the boundary conditions, agglomeration is found to occur in different fashions. The general mobility of ionic point defects in PZT is very low at room temperature and the thermally traversed distances during the cycling periods investigated in this work are in the 100 nm range. Thus, all motions of ionic defects have to occur on this scale only enhanced by the high electric fields present during fatigue.

In the center of a sample of sufficient thickness the dominant agglomeration occurs as planar agglomerates of charged oxygen vacancies in the bulk of the grains, which are directly visualized using etching techniques. These defects are very stable. They can not be removed by high electric field treatments. Only thermal heating allowing significant thermal diffusion is sufficient to disintegrate the agglomerates. These agglomerates are termed hard.

Several techniques are used to demonstrate the interaction of these etchable agglomerates with the domain system. First, the enhanced acoustic emission activity is identified to be due to the growth of defect agglomerates. It is furthermore shown that the crystal symmetry of fatigued samples very close to the morphotropic phase boundary changes from rhombohedral to tetragonal. This fact is directly correlated to the formation of planar agglomerates constituted by oxygen vacancies, which in turn induce wedge shaped domains in their immediate environment. Symmetry arguments show that an increasing interaction energy between the $\{100\}$ oriented defect planes and rhombohedral domains arises. This stored energy, or in other words, the shear stress exerted on the rhombohedral domains, finally switches the grain into the more favorably oriented tetragonal structure. For the tetragonal structure this shearing energy vanishes. The crystallographic phase change occurs despite the local boundary conditions which previously determined the crystal structure of this particular grain. In favorable samples, i.e. those that are close enough to the phase boundary, but initially all rhombohedral, the change of the entire microstructure from rhombohedral to tetragonal is seen.

The next type of agglomeration occurs in the grain boundaries in the proximity of the electrodes. Near the external electrodes, other significant modifications of the microstructure arise. In this region, the crystallographic phase switching is seen last and the number of etch grooves being the oxygen vacancy agglomerates is lowest. On the other hand, it is found, that grain boundaries change their character here. While transmission electron microscopy images of unfatigued samples did not indicate any particularities, the triple points in fatigued samples were weakened sufficiently to disappear during the ion thinning treatment for sample preparation of transmission electron microscopy. A tentative explanation of vacancy agglomeration in the grain boundaries instead of at defect planes is given. The effect is ascribed to locally lower potential differences along the grain boundaries, extending the externally prescribed potentials further into the sample. It is known, that to some extent silver from the electrodes migrates along grain boundaries and alters the conductivity. A major question arising in the context of silver diffusion is whether it is oxidized due to the high external electric fields and incorporated into the grain boundary phase as an acceptor, or whether it is still metallic. As a dopant ion more than as a metallic silver it will enhance the conductivity of the grain boundary, because the concentrations of metallic silver are well beneath the percolation limit for metallic conduction, but high as dopants. For hard dopant ions like singly charged silver, the conductivity is known to be significantly enhanced. Thus local fields will be lower in the grain boundary region and fatigue will occur later. A completely metallically conducting grain boundary is also out-ruled by the fact that grains directly at the electrode do switch their crystal structure after sufficiently many fatigue cycles and thus experience some electric field. The growth of tetragonal regions from the center of the sample towards the electrodes is shown at intermediate degrees of fatigue further supporting only partial conductivity of the grain boundaries.

The agglomeration of oxygen vacancies at grain boundaries in the sample interior is not accessible to etching techniques, because plain grain boundaries already exhibit lower etching resistance than the bulk grains. No effects are found in transmission electron microscopy. Charged point defects are very likely to also agglomerate at the grain boundary deep in the sample to some extent, but their influence is not strong enough to be identifiable by ion thinning or chemical etching.

A model for defect agglomeration itself is presented using simple rate equations. The major assumptions of the model are irreversible vacancy capture by the preexisting agglomerate and a field driven diffusion of vacancies into the agglomerate vicinity, all of which are captured. The model yields the appropriate field, temperature and frequency dependencies known from our own data and literature and is able to explain several experimental results so far not explained by the previous models. The model is compared to other existing models of fatigue that are mostly based on some space charge increase, which is actually a different effect and dominates the less sturdy forms of the fatigue phenomenon.

Space Charge, Fatigue of Less Stable Microscopic Entities:

A different mechanism is considered to be the origin of the asymmetry in the macroscopic strain. While the polarization hysteresis stays symmetric, the strain hysteresis vs. electric field (butterfly) becomes highly asymmetric, before it considerably drops in amplitude. This is ascribed to some pinning of domains by space charges developing locally. Different from the agglomerates described above, these defects are easily redistributed by high electric fields. It is shown, that the asymmetry in the strain hysteresis can be removed by applying electric fields higher than the cycling field. Under the same treatment the strain hysteresis rises back to its full amplitude, which indicates that also the hard agglomerates are surmounted, but on field reduction, the amplitude symmetrically drops back to very low values. The clamping by the hard agglomerates can be overrun by the high field, but the agglomerates themselves survive the treatment. The weakly bound defects inducing the offset polarization on the other hand are redistributed by the high fields as was found from strain data. The electronic nature of the defects is suggested, be they polarons or other localized electronic states.

Unipolar fatigue yields a very different material behavior. No formation of hard agglomerates is observed. Only offsets are induced that render the strain hysteresis asymmetric. A small offset field as well as a large offset polarization are found, but the amplitude of the electric hysteresis is hardly changed. The internal offset polarization is much more stable than in the bipolar fatigued samples. High field treatments do not re-render the strain hysteresis symmetric. The acoustic emissions in a unipolar fatigued sample do not increase for the higher fields applied to the sample. The offset polarization thus does not impede the local domain wall motion, it only superposes on the switching polarization.

At first sight, mixed loading experiments, compressive stress and synchronized unipolar sinusoidal electric external field applied to stretch the sample after the compression, yield similar results as unipolar cycling. The induced offset polarization seems to be similarly high, but upon the application of high electric fields both wings of the butterfly grow in amplitude. Also the acoustic emissions significantly increase for the higher fields, which is not the case for the simple unipolar cycling field. According to the acoustic emission results some hard agglomeration arises, but by far not as much as in the bipolar case. It thus seems that mixed loading fatigue does move ions sufficiently to transport them near some existing agglomerates, but the effect is much less pronounced than in the electric case. The movement of the ions is here more likely driven by some convection induced by the moving domain system rather than by the electric field.

Bipolar cycling is thus necessary for the rapid formation of hard agglomerates. Two aspects of bipolar switching are important with respect to this. First, the comparison with the simple model shows that the externally applied electric fields are too low to explain the observed fatigue rates. During bipolar fatigue, just before the switching of the domain system occurs, the depolarizing field of the domains and the external field add up. At this moment, significant ion motion arises, which is also supported by data from literature and explains several previously unclear findings. The second aspect is a convective effect of the moving domain system as already discussed. This is able to provide a steady flow of point defects into the respective volume around a hard agglomerate. The driving force attracting further point defects to attach to an existing agglomerate drops off with distance from the agglomerate in directions apart from the field direction, and rapidly falls beneath the necessary threshold for ion movement at some distance from the agglomerate. The agglomeration ceases if no more defects are brought into this volume. The former effect is considered more important.

Microcracking:

The formation of microcracks during fatigue is considered to be a secondary effect only, except at discontinuities in the boundary conditions. Two origins of microcracking are identified. The first one is a simple mechanical cause. At the electrode edges or similar singular points in the macroscopic boundary conditions, the electric fields and stresses diverge, or in reality become very high. Electrode edges are shown to be a source for immediate microcrack formation, which was also the case in the singular stress fields around a macroscopic crack. The second cause of microcrack formation is the increasing development of local stresses in the bulk microstructure. Microcracks are found throughout the entire sample. The stresses develop, because the domain system is clamped particularly by the hard agglomerates and can no longer accommodate the local stress due to the irregularities in the microstructure, which enhance the externally applied fields locally. In a way the effect may be termed a reduction of local toughening.

Acoustic Emission:

The acoustic emission method is employed to identify the agglomerate growth during fatigue. To be significant, an extensive study of possible sources of acoustic emissions from ferroelectrics is included. Several sources of acoustic emissions are identified, including microcracking, crystallographic phase changes and discontinuous switching. During the course of fatigue, the acoustic emission activity significantly rises, which is considered another indication of the agglomerate growth. The agglomerates stay surmountable and yield maximum acoustic emission amplitudes just before their environment is entirely tied to them and the domains completely cease to move.

A second aspect of the acoustic emission studies is a further refinement of the technique to become meaningful in the context of ferroelectrics. Single crystals of barium titanate, $BaTiO_3$, and gadolinium molybdate, $Gd_2(MoO_4)_3$, are submitted to bipolar external loads. 90° domain wall switching in $BaTiO_3$ induces acoustic emissions which are to 90% coincident with Barkhausen-pulses, the corresponding electrical current discontinuities. During the motion of a single, purely 180° domain wall in $Gd_2(MoO_4)_3$ only certain events yielded coincidence patterns. Particular dynamics of the single domain wall are assigned to the two cases.

Depoling:

Another effect, which is only briefly treated, is the fact that device performance particularly in the case of actuators can be reduced without an actual microscopic damage. This is due to bipolar or semi-bipolar driving of actuators beneath the coercive field. It induces a reduction of remnant polarization due to depoling. Different measures can be taken to reduce this effect, like acceptor doping and UV illumination during poling or hot poling, all of which stabilize the domain system of the poled state.

Of all the mechanisms discussed only microcracking is definitely fatal and can not be recovered by any device treatment except for re-sintering. Besides microcracking some of the hard agglomerates seem to resist a high temperature treatment at temperatures well above the Curie temperature, but still beneath the sintering temperature of the ceramic.

A first model for the agglomeration was developed. Domain wall dynamics will have to be incorporated into such a model to truly describe the switchable polarization. Presently, there is hardly any knowledge about the dynamics of domain walls around existing and developing agglomerates, grain boundaries or other microstructural defects. Work concerning the domain dynamics, relaxation phenomena, and the interaction of defects with domain walls is on its way, in the context of fatigue and beyond, but outside the scope of the present work.

A Solutions to Integrals

Since $V_A = \frac{\alpha}{\sqrt{\pi}} N_A^{\frac{3}{2}} b^3$ (5.27), the left part of equation (5.39) can be evaluated as follows:

$$\int_1^{N_A} \frac{V - V_A - \frac{3\alpha b^3}{2\sqrt{\pi}}\sqrt{N_A}(N-N_A)}{\gamma N_A b^2 (N-N_A)}\, dN_A$$

$$= \frac{1}{\gamma b^2} \int_1^{N_A} \frac{V + \frac{\alpha b^3}{2\sqrt{\pi}} N_A^{\frac{3}{2}} - \frac{3\alpha b^3}{2\sqrt{\pi}}\sqrt{N_A} N}{N_A(N-N_A)}\, dN_A$$

$$= \frac{V}{\gamma b^2} \underbrace{\int_1^{N_A} \frac{1}{N_A(N-N_A)}\, dN_A}_{=:I_1} + \frac{\alpha b^3}{\sqrt{\pi}\gamma b^2} \underbrace{\int_1^{N_A} \frac{\sqrt{N_A}}{N-N_A}\, dN_A}_{=:I_2}$$

$$- \frac{3\alpha b^3 N}{2\sqrt{\pi}\gamma b^2} \underbrace{\int_1^{N_A} \frac{\sqrt{N_A}}{N_A(N-N_A)}\, dN_A}_{=:I_3} .$$

For $1 \leq N_A < N$ and large N_A, which is inherently the case, the integrals I_1, I_2 and I_3 can be evaluated as (see, for example, Bronstein and Semendjajew, 1985):

$$I_1 = \left[-\frac{1}{N} \ln \frac{N-N_A}{N_A} \right]_1^{N_A} = -\frac{1}{N}\left(\ln \frac{N-N_A}{N_A} - \ln(N-1) \right) \approx -\frac{1}{N} \ln \frac{N-N_A}{N_A}$$

$$I_2 = \left[-2\sqrt{N_A} + \frac{1}{\sqrt{N}} \ln \frac{\sqrt{N}+\sqrt{N_A}}{\sqrt{N}-\sqrt{N_A}} \right]_1^{N_A} \approx \frac{1}{\sqrt{N}} \ln \frac{\sqrt{N}-\sqrt{N_A}}{\sqrt{N}+\sqrt{N_A}} - 2\sqrt{N_A}$$

$$I_3 = \left[\frac{1}{\sqrt{N}} \ln \frac{\sqrt{N}+\sqrt{N_A}}{\sqrt{N}-\sqrt{N_A}} \right]_1^{N_A} \approx \frac{1}{\sqrt{N}} \ln \frac{\sqrt{N}-\sqrt{N_A}}{\sqrt{N}+\sqrt{N_A}} .$$

Hence (5.39) can be rewritten as

$$n \approx -\frac{V}{\gamma b^2 N} \ln \frac{N-N_A}{N_A} + \frac{\alpha b^3}{\sqrt{\pi}\gamma b^2}\left(\frac{1}{\sqrt{N}} \ln \frac{\sqrt{N}-\sqrt{N_A}}{\sqrt{N}+\sqrt{N_A}} - 2\sqrt{N_A}\right)$$

$$+ \frac{3\alpha b^3 N}{2\sqrt{\pi}\gamma b^2}\left(\frac{1}{\sqrt{N}} \ln \frac{\sqrt{N}-\sqrt{N_A}}{\sqrt{N}+\sqrt{N_A}}\right)$$

$$\approx \frac{-1}{\gamma b^2 \rho_o} \ln \frac{N-N_A}{N_A} + \frac{3\alpha b}{2\sqrt{\pi}\gamma} N \ln \frac{\sqrt{N}-\sqrt{N_A}}{\sqrt{N}+\sqrt{N_A}} - 2\frac{\alpha b}{\sqrt{\pi}\gamma}\sqrt{N_A} \ ,$$

where the lower integration boundaries were dropped due to their small absolute values.

References

Aburatani, H. and Uchino, K. (1996). Acoustic emission (AE) measurement technique in piezoelectric ceramics. *Jpn. J. Appl. Phys.*, **35**, L516–L518.

Aburatani, H., Yoshikawa, S., Uchino, K., and Vries, J. W. C. de (1998). A study of acoustic emission in piezoelectric multilayer ceramic actuators. *Jpn. J. Appl. Phys.*, **37**, 204–209.

Aki, K. and Richards, P. G. (1980). *Quantitative Seismology, Theory and Methods.* W.H. Freeman and Company, SF, USA.

Al-Shareef, H. N., Tuttle, B. A., Warren, W. L., Headley, T. J., Dimos, D., Voigt, J. A., and Nasby, R. D. (1996a). Effect of B-site cation stochiometry on electrical fatigue of $RuO_2/Pb(Zr_xTi_{1-x})O_3/RuO_2$ capacitors. *J. Appl. Phys.*, **79**, 1013–1016.

Al-Shareef, H. N., Tuttle, B. A., Warren, W. L., Dimos, D., Raymond, M. V., and Rodriguez, M. A. (1996b). Low temperature processing of Nb-doped $Pb(Zr,Ti)O_3$ capacitors with $La_{0.5}Sr_{0.5}CoO_3$ electrodes. *Appl. Phys. Lett.*, **68**, 272–74.

Al-Shareef, H. N., Dimos, D., Boyle, T. J., Warren, W. L., and Tuttle, B. A. (1996c). Qualitative model for the fatigue-free behavior of $SrBi_2Ta_2O_9$. *Appl. Phys. Lett.*, **68**, 690–92.

Al-Shareef, H. N., Dimos, D., Warren, W. L., and Tuttle, B. A. (1997). A model for optical and electrical polarization fatigue in $SrBi_2Ta_2O_9$ and $Pb(Zr,Ti)O_3$. *Integrated Ferroelectrics*, **15**, 53–67.

Allnatt, A. R. and Lidiard, A. B. (1993). *Atomic Transport in Solids.* Cambridge University Press.

Anderson, J. R., Brady, G. W., Merz, W. J., and Remeika, J. P. (1955). Effects of ambient atmosphere on the stability of barium titanate. *J. Appl. Phys.*, **26**, 1387–1388.

Arai, M., Sugawara, Y., and Uchino, K. (1990). AE measurements in ferroelectrics. *IEEE Ultrasonics Symposium Proc.*, **3**, 1197–1200.

Arlt, G. (1990a). The influence of microstructure on the properties of ferroelectric ceramics. *Ferroelectrics*, **104**, 217–227.

Arlt, G. (1990b). Review: Twinning in ferroelectric and ferroelastic ceramics: Stress relief. *J. Mater. Sci.*, **25**, 2655–2666.

Arlt, G. and Neumann, H. (1988). Internal bias in ferroelectric ceramics: Origin and time dependence. *Ferroelectrics*, **87**, 109–120.

Arlt, G. and Sasko, P. (1980). Domain configuration and equilibrium size of domains in $BaTiO_3$ ceramics. *J. Appl. Phys.*, **51**, 4956–4960.

Arlt, G., Böttger, U., and Witte, S. (1995). Dielectric dispersion of ferroelectric ceramics and single crystals by sound generation in piezoelectric domains. *J. Am. Ceram. Soc.*, **78**, 1097–1100.

Avrami, M. (1939). Kinetics of phase change. I. *J. Chem. Phys.*, **7**, 1103–1112.

Avrami, M. (1940). Kinetics of phase change. II, transformation-time relations for random distribution of nuclei. *J. Chem. Phys.*, **8**, 212–224.

Avrami, M. (1941). Granulation, phase change, and microstructure, kinetics of phase change. III. *J. Chem. Phys.*, **9**, 177–184.

Baiatu, T., Waser, R., and Härdtl, K. H. (1990). D.c. electrical degradation of perovskite-type titanates: III, A model of the mechanism. *J. Am. Ceram. Soc.*, **73**, 1663–1673.

Bao, D., Wakiya, N., Shinozaki, K., and Mizutani, N. (2002a). Ferroelectric properties of sandwich structures $(Bi,La)_4Ti_3O_{12}/Pb(Zr,Ti)O_3/(Bi,La)_4Ti_3O_{12}$ thin films on multilayered $Pt/Ti/SiO_2/Si$ substrates. *J. Phys. D: Appl. Phys.*, **35**, L1–L5.

Bao, D., Wakiya, N., Shinozaki, K., and Mizutani, N. (2002b). Preparation and electrical properties of $(Bi,La)_4Ti_3O_{12}/Pb(Zr,Ti)O_3/(Bi,La)_4Ti_3O_{12}$ multilayer thin films by a chemical solution deposition. *Ferroelectrics*, **270**, 27–32.

Barkley, J. R., Brixner, L. H., Hogan, E. M., and Waring, Jr., R. K. (1972). Control and application of domain wall motion in gadolinium molybdate. *Ferroelectrics*, **3**, 191–197.

Beattie, A. G. (1976). Energy analysis in acoustic emission. *Materials Evaluation*, **April**, 73–78.

Beattie, A. G. (1983). Acoustic emission, principles and instrumentation. *J. Acoustic Emission*, **2 (1/2)**, 95–128.

Becerro, A. I., McCammon, C., Langenhorst, F., Seifert, F., and Angel, R. (1999). Oxygen vacancy ordering in $CaTiO_3$-$CaFeO_{2.5}$ perovskites: From isolated defects to infinite sheets. *Phase Transitions*, **69**, 133–146.

Bernstein, S. D., Kisler, Y., Wahl, J. M., Bernacki, S. E., and Collins, S. R. (1992). Effects of stoichiometry on PZT thin film capacitor properties. *Mater. Res. Soc. Symp. Proc.*, **243**, 373–378.

Bernstein, S. D., Wong, T. Y., Kisler, Y., and Tustison, R. W. (1993). Fatigue of ferroelectric $PbZr_xTi_yO_3$ capacitors with Ru and RuO_x electrodes. *J. Mater. Res.*, **8**, 12–13.

Bobnar, V., Kutnjak, Z., Levstik, A., Holc, J., Kosec, M., Hauke, T., Steinhausen, R., and Beige, H. (1999). Correlation between fatigue and piezoelectric properties in $(Pb,La)(Zr,Ti)O_3$ thick films. *J. Appl. Phys.*, **85**, 622–624.

Bornand, V. and Trolier-McKinstry, S. (2000). Structural and dielectric properties of pulsed laser deposited $Pb[Yb_{1/2}Nb_{1/2}]O_3$-$PbTiO_3$ thin films. *Mater. Res. Soc. Symp. Proc.*, **603**, 143–148.

Bornard, V., Trolier-McKinstry, S., Takemura, K., and Randall, C. A. (2000). Orientation dependence of fatigue behavior in relaxor ferroelectric-$PbTiO_3$ thin films. *J. Appl. Phys.*, **87**, 3965–3972.

Bratkovsky, A. M. and Levanyuk, A. P. (2000a). Abrupt appearance of the domain pattern and fatigue of thin ferroelectric films. *Phys. Rev. Lett.*, **84**, 3177–80.

Bratkovsky, A. M. and Levanyuk, A. P. (2000b). Ferroelectric phase transitions in films with depletion charge. *Phys. Rev. B*, **61**, 15042–15050.

Brazier, M., Mansour, S., and McElfresh, M. (1999). Ferroelectric fatigue of $Pb(Zr,Ti)O_3$ thin films measured in atmospheres of varying oxygen concentration. *Appl. Phys. Lett.*, **74**, 4032–4033.

Brennan, C. (1993). Model of ferroelectric fatigue due to defect/domain interactions. *Ferroelectrics*, **150**, 199–208.

Brennan, C. J. (1995). Defect chemistry model of the ferroelectric-electrode interface. *Integrated Ferroelectrics*, **7**, 93–109.

Brennan, C. J., Parrella, R. D., and Larsen, D. E. (1994). Temperature dependent fatigue rates in thin- film ferroelectric capacitors. *Ferroelectrics*, **151**, 33–38.

Bronstein, I. N. and Semendjajew, K. A. (1985). *Taschenbuch der Mathematik.* Teubner, Leipzig.

Bunkin, A. Yu. and Nishnevitch, M. B. (1995). Interaction of fluid flow under a rotating crystal with the growing surface. *J. Cryst. Growth*, **156**, 454–458.

Burridge, R. and Knopoff, L. (1964). Body force equivalents for seismic dislocations. *Bull. Seismol. Soc. Am.*, **54**, 1875–1888.

Bursill, L. A., Reaney, I. M., Vijay, D. P., and Desu, S. B. (1994). Comparison of lead zirconate titanate thin films on ruthenium oxide and platinum electrodes. *J. Appl. Phys.*, **75**, 1521–1525.

Buscaglia, M. T., Buscaglia, V., Viviani, M., and Nanni, P. (2001). Atomistic simulation of dopant incorporation in barium titanate. *J. Am. Ceram. Soc.*, **84**, 376–384.

Cain, M. G. and Stewart, M. nad Gee, M. (1999). Degradation of piezoelectric materials. *Nat. Phys. Lab. Rep. (UK), CMMT(A)*, **148**, 1–39.

Cao, H. and Evans, A. G. (1993). Nonlinear deformation of ferroelectric ceramics. *J. Am. Ceram. Soc.*, **76**, 890–896.

Carl, K. (1975). Ferroelectric properties and fatiguing effects of modified $PbTiO_3$ ceramics. *Ferroelectrics*, **9**, 23–32.

Chan, N.-H., Sharma, R. K., and Smyth, D. M. (1981). Nonstoichiometry in undoped $BaTiO_3$. *J. Am. Ceram. Soc.*, **64**, 556–562.

Chan, N. H., Sharma, R. K., and Smyth, D. M. (1982). Nonstoichiometry in acceptor-doped $BaTiO_3$. *J. Am. Ceram. Soc.*, **65**, 167–170.

Chao, Gwo-Chin and Wu, Jenn-Ming (2001). Leakage current and fatigue properties of $Pb(Zr,Ti)O_3$ ferroelectric films prepared by rf-magnetron sputtering on textured $LaNiO_3$ electrode. *Jpn. J. Appl. Phys.*, **40**, 2417–2422.

Chen, Chin-Yi and Tuan, Wei-Hsing (2000). Effect of silver on the sintering and grain-growth behavior of barium titanate. *J. Am. Ceram. Soc.*, **83**, 2988–2992.

Chen, J., Harmer, M. P., and Smyth, D. M. (1994a). Compositional control of ferroelectric fatigue in perovskite ferroelectric ceramics and thin films. *J. Appl. Phys.*, **76**, 5394–5398.

Chen, W., Lupascu, D. C., Rödel, J., and Lynch, C. S. (2001). Short crack R-curves of ferroelectric and electrostrictive PLZT. *J. Am. Ceram. Soc.*, **84**, 593–597.

Chen, X., Kingon, A. I., Al-Shreef, H., and Bellur, K. R. (1994b). Electrical transport and dielectric breakdown in $Pb(Zr,Ti)O_3$ thin films. *Ferroelectrics*, **151**, 133–138.

Chen, Xi and Fang, Dai-Ning (1998). A nonlinear constitutive theory for ferroelectrics. *Key Engineering Materials*, **145-149**, 977–982.

Cheng, Hsiu-Fung, Chen, Yi-Chun, Chou, Chen-Chia, Chang, Kuang-Chung, Hou, Chun-Shu, and Lin, I-Nan (2000). Comparison on the effect of $(La_{0.5}Sr_{0.5})MnO_3$

and $(La_{0.5}Sr_{0.5})CoO_3$ buffer layers on fatigue properties of $(Pb_{0.6}Sr_{0.4})TiO_3$ thin films prepared by pulsed laser deposition. *J. Appl. Phys.*, **87**, 8695–8699.

Cho, Seong Moon and Jeon, Duk Young (1999). Effect of annealing conditions on the leakage current characteristics of ferroelectric PZT thin films grown by sol-gel process. *Thin Solid Films*, **338**, 149–154.

Choi, D. G. and Choi, S. K. (1997). Dynamic behaviour of domains during poling by acoustic emission measurements in La-modified $PbTiO_3$ ferroelectric ceramics. *J. Mater. Sci.*, **32**, 421–425.

Chon, Uong, Jang, H. M., Kim, M. G., and Chang, C. H. (2002). Layered perovskites with giant spontaneous polarizations for nonvolatile memories. *Phys. Rev. Lett.*, **89**, 087601–1–4.

Chou, Chen-Chia, Hou, Chun-Shu, and Pan, Han-Chang (2001). Domain boundary pinning and nucleation of ferroelectric $(Pb_{1-x}Sr_x)TiO_3$ ceramics. *Ferroelectrics*, **261**, 185–190.

Chu, Fan and Fox, G. (2001). Relationship between Pb content, crystallographic texture and ferroelectric properties of PLZT thin films for FRAM applications. *Integrated Ferroelectrics*, **33**, 19–26.

Chynoweth, A. G. (1958). Barkhausen pulses in barium titanate. *Phys. Rev.*, **110**, 1316–1332.

Chynoweth, A. G. (1959). Radiation damage effects in ferroelectric triglycine sulfate. *Phys. Rev.*, **113**, 159–166.

Cillesen, J. F. M., Prins, M. W. J., and Wolf, R. M. (1997). Thickness dependence of the switching voltage in all-oxide ferroelectric thin-film capacitors prepared by pulsed laser deposition. *J. Appl. Phys.*, **81**, 2777–2783.

Clarke, D. R. and Aurora, A. (1983). Acoustic emission characterization of the tetragonal-monoclinic phase transformation in zirconia. *Advances in Ceramics, Proc. 2nd Int. Conf. Sci. Techn. Zirconia, N. Claussen, M. Rühle, A.H. Heuer Eds.*, **12**, 54–63.

Colla, E. L., Kholkin, A. L., Taylor, D., Tagantsev, A. K., Brooks, K. G., and Setter, N. (1995). Characterisation of the fatigued state of ferroelectric PZT thin-film capacitors. *Microelectronic Engin.*, **29**, 145–148.

Colla, E. L., Tagantsev, A. K., Taylor, D. V., and Kholkin, A. L. (1997). Fatigued state of Pt-{PZT}-PT system. *Integrated Ferroelectrics*, **18**, 19–28.

Colla, E. L., Hong, Seungbum, Taylor, D. V., Tagantsev, A. K., and Setter, N. (1998a). Direct observation of region by region suppression of the switchable polarization (fatigue) in $Pb(Zr,Ti)O_3$ thin film capacitors with Pt electrodes. *Appl. Phys. Lett.*, **72**, 2763–2765.

Colla, E. L., Taylor, D. V., Tagantsev, A. K., and Setter, N. (1998b). Discrimination between bulk and interface scenarios for the suppression of the switchable polarization (fatigue) in $Pb(Zr,Ti)O_3$ thin films capacitors with Pt electrodes. *Appl. Phys. Lett.*, **72**, 2478–2480.

Colla, E. L., Tagantsev, A. K., Taylor, D. V., and Kholkin, A. L. (1998c). Field self-adjusted suppression of the switching polarisation in ferroelectric PZT thin films with Pt-electrodes. *J. Korean Phys. Soc.*, **32**, 1353–1356.

Cross, L. E. (1994). Relaxor ferroelectrics: An overview. *Ferroelectrics*, **151**, 305–320.

Damjanovic, D. (1998). Ferroelectric, dielectric and piezoelectric properties of ferroelectric thin films and ceramics. *Rep. Prog. Phys.*, **61**, 1267–1324.

Daniels, J. (1976). Part II. Defect equilibria in acceptor doped barium titanate. *Philips Res. Repts.*, **31**, 505–515.

Daniels, J. and Härdtl, K. H. (1976). Part I. Electrical conductivity at high temperatures of donor-doped barium titanate ceramics. *Philips Res. Repts.*, **31**, 489–504.

Daniels, J. and Wernicke, R. (1976). Part V. New aspects of an improved PTC model. *Philips Res. Repts.*, **31**, 544–559.

Dausch, D. E. (1997). Ferroelectric polarization fatigue in PZT-based RAINBOWs and bulk ceramics. *J. Am. Ceram. Soc.*, **80**, 2355–2360.

Dawber, M. and Scott, J. (2000a). A model for fatigue in ferroelectric perovskite thin films. *Appl. Phys. Lett.*, **76**, 1060–1062.

Dawber, M. and Scott, J. (2000b). A model for fatigue in ferroelectric perovskite thin films. *Appl. Phys. Lett.*, **76**, 3655.

Dawber, M. and Scott, J. F. (2001). Fatigue and oxygen vacancy ordering in thin-film and bulk single crystal ferroelectrics. *Integrated Ferroelectrics*, **32**, 259–266.

Dawber, M. and Scott, J.F. (2002). Models of electrode-dielectric interfaces in ferroelectric thin-film devices. *subm.*, pages 1–18.

Dawber, M., Rios, S., Scott, J. F., Zhang, Qi, and Whatmore, R. W. (2001). Cryogenic electrical studies of manganese doped lead scandium tantalate thin films: Phase transitions or domain wall dynamics? In *Fundamental Physics of Ferroelectrics, H. Krakauer, edt., Amer. Inst. Phys. CP582*, pages 1–10.

Dawber, M., Sinnamon, L. J., Scott, J. F., and Gregg, J. M. (2002). Electrode field penetration: A new interpretation of tunneling currents in barium strontium titanate (BST) thin films. *Ferroelectrics*, **268**, 35–40.

Denk, I., Münch, W., and Maier, J. (1995). Partial conductivities in $SrTiO_3$: Bulk polarization experiments, oxygen concentration cell measurements, and defect-chemical modeling. *J. Am. Ceram. Soc.*, **78**, 3265–3272.

Denk, I., Noll, F., and Maier, J. (1997). In situ profiles of oxygen diffusion in $SrTiO_3$: Bulk behaviour and boundary effects. *J. Am. Ceram. Soc.*, **80**, 279–285.

Desu, S. B. (1995). Minimization of fatigue in ferroelectric films. *phys. stat. sol. (a)*, **151**, 467–481.

Desu, S. B. and Vijay, D. P. (1995). Novel fatigue-free layered structure ferroelectric thin films. *Mater. Sci. & Engin. B*, **32**, 75–81.

Desu, S. B. and Yoo, In-K. (1993a). Electrochemical models of failure in oxide perovskites. *Integrated Ferroelectrics*, **3**, 365–376.

Desu, S. B. and Yoo, In-K. (1993b). Time dependent dielectric breakdown in $BaTiO_3$ thin films. *J. Electrochem. Soc.*, **140**, 133–35.

Devonshire, A. F. (1954). Theory of ferroelectrics. *Advances in Physics*, **3**, 85–129.

Dimos, D., Warren, W. L., and Tuttle, B. A. (1993). Photo-induced and electrooptic properties of (Pb,La)(Zr,Ti)O_3 films. *Mater. Res. Soc. Symp. Proc.*, **310**, 87–98.

Dimos, D., Warren, W.L., Sinclair, M.B., Tuttle, B.A., and Schwartz, R.W. (1994). Photoinduced hysteresis changes and optical storage in (Pb,La)(Zr,Ti)O_3 thin films and ceramics. *J. Appl. Phys.*, **76**, 4305–4315.

Dimos, D., Al-Shareef, H. N., Warren, W.L., and Tuttle, B.A. (1996). Photoinduced changes in the fatigue behavior of $SrBi_2Ta_2O_9$ and Pb(Zr,Ti)O_3 thin films. *J. Appl. Phys.*, **80**, 1682–1687.

Ding, Y., Liu, J. S., Maclaren, I., Wang, Y. N., and Kuo, K. H. (2001a). Study of domain walls and their effect on switching property in Pb(Zr,Ti)O_3, $SrBi_2Ta_2O_9$ and $Bi_4Ti_3O_{12}$. *Ferroelectrics*, **262**, 37–46.

Ding, Y., Liu, J. S., Qin, H. X., Zhu, J. S., and Wang, Y. N. (2001b). Why lanthanum-substituted bismuth titanate becomes fatigue free in a ferroelectric capacitor with platinum electrodes. *Appl. Phys. Lett.*, **78**, 4175–4177.

Ding, Yong, Liu, Jianshe, Wang, Yening, and Kuo, Kehsin (2001c). Ferroelectric domain structure, domain wall mobility, and related fatigue-free behavior in $SrBi_2Ta_2O_9$. *Ferroelectrics*, **251**, 165–174.

Du, Xiao-Hong, Zheng, Jiehui, Belegundu, U., and Uchino, K. (1998). Crystal orientation dependence of piezoelectric properties of lead zirconate titanate near the morphotropic phase boundary. *Appl. Phys. Lett.*, **72**, 2421–2423.

Du, Xiaofeng and Chen, I-Wei (1998a). Fatigue of $Pb(Zr_{0.58}Ti_{0.47})O_3$ ferroelectric thin films. *J. Appl. Phys.*, **83**, 7789–7798.

Du, Xiaofeng and Chen, I-Wei (1998b). Model experiments on fatigue of $Pb(Zr_{0.53}Ti_{0.47})O_3$ ferroelectric thin films. *Appl. Phys. Lett.*, **72**, 1923–1925.

Duiker, H. M. and Beale, P. D. (1990). Grain-size effect in ferroelectric switching. *Phys. Rev. B*, **41**, 490–495.

Duiker, H. M., Beale, P. D., Scott, J. F., Araujo, C. A. Paz de, Melnick, B. M., Cuchiaro, J. D., and McMillan, L. D. (1990). Fatigue and switching in ferroelectric memories: Theory and experiment. *J. Appl. Phys.*, **68**, 5783–5791.

Dul'kin, E. A. (1999). Ferroic phase boundaries investigations by the acoustic emission method. *Mater. Res. Innovation*, **2**, 338–340.

Dul'kin, E. A., Gavrilyachenko, V. G., and Semenchev, A. F. (1992). Investigation of sound emission from $BaTiO_3$-type ferroelectric crystals near their phase transition. *Sov. Phys. Solid State*, **34**, 863–864.

Dul'kin, E. A., Gavrilyachenko, V. G., and Semenchev, A. F. (1993). Acoustic emission due to a phase transition in barium titanate crystals subjected to sequential etching. *Sov. Phys. Solid State*, **35**, 1016–1017.

Dunegan, H. L. and Hartman, W. F. (1981). *Advances in Acoustic Emission, Proc. Int. Conf. On AE, Anaheim, CA, 1979, eds.* Dunhart.

Eror, N. G. and Smyth, D. M. (1978). Nonstoichiometric disorder in single-crystalline $BaTiO_3$ at elevated temperatures. *J. Solid State Chem.*, **24**, 235–244.

Fatuzzo, E. (1962). Theoretical considerations on the switching transient in ferroelectrics. *Phys. Rev.*, **127**, 1999–2005.

Fatuzzo, E. and Merz, W. J. (1967). *Ferroelectricity.* North-Holland Publishing Company, Amsterdam.

Fernandez, J. F., Duran, P., and Moure, C. (1990). Dielectric and piezoelectric aging of pure and Nb- doped $BaTiO_3$ ceramics. *Ferroelectrics*, **106**, 381–386.

Fett, T., Munz, D., and Thun, G. (1998). Nonsymmetric deformation behaviour of PZT determined in bending tests. *J. Am. Ceram. Soc.*, **81**, 269–72.

Flippen, R. B. (1975). Domain wall dynamics in ferroelectric/ferroelastic molybdates. *J. Appl. Phys.*, **46**, 1068–1071.

Fraser, D. B. and Maldonado, J. R. (1970). Improved aging and switching of lead zirconate-titanate ceramics with indium electrodes. *J. Appl. Phys.*, **41**, 2172–2176.

Freiman, S. W. and Pohanka, R. C. (1989). Review of mechanically related failures of ceramic capacitors and capacitor materials. *J. Am. Ceram. Soc.*, **72**, 2258–2263.

Fridkin, V. (1980). *Ferroelectric Semiconductors.* Consultants Bureau, Plenum Publishing Corporation, New York.

Friessnegg, T., Aggarwal, S., Nielsen, B., Ramesh, R., Keeble, J. D., and Poindexter, H. E. (2000). A study of vacancy-related defects in (Pb,La)(Zr,Ti)O_3 thin films using positron annihilation. *IEEE*, **47**, 916–920.

Friessnegg, T., Nielsen, B., and Keeble, D. J. (2001). Detection of oxygen vacancies in (Pb,La)(Zr,Ti)O_3 thin film capacitors using positron annihilation. *Integrated Ferroelectrics*, **32**, 179–197.

Furuta, A. and Uchino, K. (1993). Dynamic observation of crack propagation in piezoelectric multilayer actuators. *J. Am. Ceram. Soc.*, **76**, 1615–1617.

Furuta, A. and Uchino, K. (1994). Destruction mechanism of multilayer ceramic actuators: Case of antiferroelectrics. *Ferroelectrics*, **160**, 277–285.

Gondro, E., Kühn, C., Schuler, F., and Kowarik, O. (2000). Physics based fatigue compact model for ferroelectric capacitors. In *IEEE Proc. 12th Intern. Symp. on the Appl. of Ferroelectrics, ISAF*, pages 615–618.

Grabec, I. and Esmail, E. (1986). Analysis of acoustic emission during martensitic transformation of CuZnAl alloy by measurement of forces. *Phys. Lett.*, **113A**, 376–378.

Grenier, J. C., Pouchard, M., and Hagenmuller, P. (1981). Vacancy ordering in oxygen-deficient perovskite-related ferrites. *Structure and Bonding, Springer, Berlin*, **47**, 2–25.

Grenier, P., Jandl, S., Blouin, M., and Boatner, L. A. (1992). Study of ferroelectric microdomains due to oxygen vacancies in $KTaO_3$. *Ferroelectrics*, **137**, 105–111.

Grindlay, J. (1970). *An Introduction to the Phenomenological Theory of Ferroelectrics.* Pergamon Press, Oxford.

Groot, S. R. de and Mazur, P. (1962). *Non-Equilibrium Thermodynamics.* North Holland Publishing, Amsterdam.

Grossmann, M., Lohse, O., Bolten, D., Boettger, U., Schneller, T., and Waser, R. (2002a). The interface screening model as origin of imprint in pbzr$_x$ti$_{1-x}$o$_3$ thin films. i. dopant, illumination, and bias dependence. *J. Appl. Phys.*, **92**, 2680–2687.

Grossmann, M., Lohse, O., Bolten, D., Boettger, U., and Waser, R. (2002b). The interface screening model as origin of imprint in pbzr$_x$ti$_{1-x}$o$_3$ thin films. ii. numerical simulation and verification. *J. Appl. Phys.*, **92**, 2688–2696.

Gruverman, A., Auciello, O., and Tokumoto, H. (1996a). Nanoscale investigation of fatigue effect in Pb(Zr,Ti)O_3 films. *Appl. Phys. Lett.*, **69**, 3191–3193.

Gruverman, A., Auciello, O., Hatano, J., and Tokumoto, H. (1996b). Scanning force microscopy as a tool for nanoscale study of ferrolectric domains. *Ferroelectrics*, **184**, 11–20.

Gruverman, A., Auciello, O., and Tokumoto, H. (1998a). Imaging and control of domain structures in ferroelectric thin films via scanning force microscopy. *Ann. Rev. Mater. Sci.*, **28**, 101–123.

Gruverman, A., Auciello, O., and Tokumoto, H. (1998b). Scanning force microscopy: Application to nanoscale studies of ferroelectric domains. *Integrated Ferroelectrics*, **19**, 49–83.

Güttler, B., Bismayer, U., Groves, P., and Salje, E. (1995). Fatigue mechanisms in thin film PZT memory materials. *Semicond. Sci. Technol.*, **10**, 245–248.

Haertling, G. H. (1987). PLZT electrooptic materials and applications - a review. *Ferroelectrics*, **75**, 25–55.

Haertling, G. H. (1994). Rainbow ceramics - a new type of ultra-high displacement actuator. *Am. Cer. Soc. Bull.*, **73**, 93–96.

Hafid, M., Handerek, J., Kugel, G. E., and Fontana, M. D. (1992). Raman spectroscopy investigations of Nb-doped lead zirconate titanate ceramics with AFE-FE-PE phase sequence. *Ferroelectrics*, **125**, 477–482.

Hagemann, H.-J. (1980). *Akzeptorionen in* $BaTiO_3$ *und* $SrTiO_3$ *und ihre Auswirkung auf die Eigenschaften von Titanatkeramiken.* Dissertation, Aachen University of Technology, Germany.

Hagenbeck, R., Schneider-Störmann, L., Vollmann, M., and Waser, R. (1996). Numerical simulation of the defect chemistry and electrostatics at grain boundaries in titanate ceramics. *Mater. Sci. & Engin. B*, **39**, 179–187.

Hammer, M. (1996). *Herstellung und Gefüge-Eigenschaftskorrelationen von PZT-Keramiken.* PhD-Thesis, University Karlsruhe, Germany.

Hammer, M., Monty, C., Endriss, A., and Hoffmann, M. J. (1998). Correlation between surface texture and chemical composition in undoped, hard and soft piezoelectric PZT ceramics. *J. Am. Ceram. Soc.*, **81**, 721–724.

Hammer, M., Endriss, A., Lupascu, D. C., and Hoffmann, M. J. (1999). Influence of microstructure on microscopic and macroscopic strain behaviour of soft PZT ceramics. In *C. Galassi et al. (eds.), Piezoelectric Materials: Advances in Science, Technology and Applications, 137-147, (NATO-Workshop, 24.-28.5.1999, Bukarest)*, pages 137–147.

Han, Geunjo and Lee, J. (2000). The effect of metallic oxide layer on reliability of lead zirconate titanate thin film capacitors. *Surf. Sci. Technol.*, **131**, 543–547.

Hase, T., Noguchi, T., Takemura, K., and Miyasaka, Y. (1998). Fatigue characteristics of PZT capacitors with Ir/IrO_x electrodes. In *Proc. 11th IEEE Int. Symp. Appl. Ferroel.*, pages 7–10.

Hatano, H. (1977). Acoustic emission and stacking fault energy. *J. Appl. Phys.*, **48**, 4397–4399.

Hauke, T., Beige, H., Giersbach, M., Seifert, S., and Sporn, D. (2000). Influence of cyclic mechanical and electrical load on the properties of PZT(53/47) films on metallic substrates. In *IEEE Proc. 12th Intern. Symp. on the Appl. of Ferroelectrics, ISAF*, pages 897–900.

Hauke, T., Beige, H., Giersbach, M., Seifert, S., and Sporn, D. (2001). Mechanical and electric fatigue of PZT(53/47) films on metallic substrates. *Integrated Ferroelectrics*, **35**, 219–228.

Heiple, C. R., Carpenter, S. H., and Carr, M. J. (1981). Acoustic emission from dislocation motion in precipitation - strengthened alloys. *Metal Sci.*, **15**, 587–598.

Helke, G. and Kirsch, W. (1971). Dielektrische und Piezoelektrische Eigenschaften der ternären keramischen festen Lösungen $Pb(Ni_{1/3}Sb_{2/3})O_3$-$PbTiO_3$-$PbZrO_3$. *Hermsdorfer Technische Mitteilungen*, **32**, 1010–1015.

Hellwege, K. H. (1988). *Einführung in die Festkörperphysik.* Springer-Verlag, Berlin.

Hench, L. L. and West, J. K. (1990). *Principles of Electronic Ceramics.* John Wiley & Sons, New York.

Hennings, D. (1976). Part III. Thermogravimetric investigation. *Philips Res. Repts.*, **31**, 516–525.

Hill, M. D., White, G. S., Hwang, C. S., and Lloyd, I. K. (1996). Cyclic damage in lead zirconate titanate. *J. Am. Ceram. Soc.*, **79**, 1915–1920.

Höchli, U. T. (1972). Elastic constants and soft optical modes in gadolinium molybdate. *Phys. Rev. B*, **6**, 1814–1823.

Hoffmann, M. J., Hammer, M., Endriss, A., and Lupascu, D. C. (2001). Correlation between microstructure, strain behavior, and acoustic emission of soft PZT ceramics. *Acta mater.*, **49**, 1301–1310.

Hull, D. and Bacon, D. J. (1984). *Introduction to Dislocations.* Pergamon Press, Oxford.

Hwang, Sun, Rho, Jun Seo, Chang, Young Chul, and Chang, Ho Jung (2001). Preparation and characterization of (Bi,La)Ti_3O_{12} thin films for nonvolatile memory. In *Adv. Electron. Mater. and Packaging 2001 (Cat. No. 01EX506) Piscataway IEEE 2001*, pages 188–192.

Ishibashi, Y. (1985). Polarization reversal kinetics in ferroelectric liquid crystals. *Jpn. J. Appl. Phys., suppl.*, **24-3**, 126–129.

Ishibashi, Y. and Takagi, Y. (1971). Note on ferroelectric domain switching. *J. Phys. Soc. Jpn.*, **31**, 506–510.

Jackson, J. D. (1975). *Classical Electrodynamics.* John Wiley & Sons, New York.

Jaffe, B., Cook, Jr., W. R., and Jaffe, H. (1971). *Piezoelectric Ceramics.* Academic Press, Marietta, OH.

Jang, Hyuk Kyoo, Lee, Cheol Eui, and Noh, Seung Jeong (2000). Electrical fatigues in oxygen plasma-treated sol-gel-derived lead-zirconate-titanate films. *Jpn. J. Appl. Phys., Pt. 2*, **39**, L1105–L1107.

Jiang, Q. and Cross, L. E. (1993). Effect of porosity on electrical fatigue behavior in PLZT and PZT ceramics. *J. Mater. Sci.*, **28**, 4536–4543.

Jiang, Q., Subbarao, E. C., and Cross, L. E. (1994a). Effects of electrodes and electroding methods on fatigue behavior in ferroelectric materials. *Ferroelectrics*, **154**, 119–124.

Jiang, Q., Cao, W., and Cross, L. E. (1994b). Electric fatigue in lead zirconate titanate ceramics. *J. Am. Ceram. Soc.*, **77**, 211–215.

Jiang, Q., Subbarao, E. C., and Cross, L. E. (1994c). Grain size dependence of electric fatigue behavior of hot pressed PLZT ferroelectric ceramics. *Acta metall. mater.*, **42**, 3687–3694.

Jiang, Q. Y., Cao, Wenwu, and Cross, L. E. (1992). Electric fatigue initiated by surface contamination in high polarization ceramics. In *IEEE Proc. 8th Int. Symp. Appl. Ferroelectrics*, pages 107–110.

Jiang, Q. Y., Subbarao, E. C., and Cross, L. E. (1994d). Effect of composition and temperature on electric fatigue of La-doped lead zirconate titanate ceramics. *J. Appl. Phys.*, **75**, 7433–7443.

Jiang, Q. Y., Subbarao, E. C., and Cross, L. E. (1994e). Fatigue in PLZT: Acoustic emission as a discriminator between microcracking and domain switching. *Ferroelectrics*, **154**, 113–118.

Jin, Shi-Zhao, Lim, Ji Eun, Cho, Moon Joo, Hwang, Cheol Seong, and Kim, Seung-Hyun (2002). Heat-treatment-induced ferroelectric fatigue of Pt/$Sr_{1-x}Bi_{2+y}Ta_2O_9$/Pt thin-film capacitors. *Appl. Phys. Lett.*, **81**, 1477–1479.

Jonker, G. H. (1972). Nature of aging in ferroelectric ceramics. *J. Am. Ceram. Soc.*, **55**, 57–58.

Kala, T. (1991). Electronic properties of Pb(Zr,Ti)O_3 solid solutions. *Phase Transitions*, **36**, 65–88.

Kalitenko, V. A., Perga, V. M., and Salivonov, I. N. (1980). Determination of the temperatures of phase transitions in barium and strontium titanates by acoustic emission. *Sov. Phys. Solid State*, **22**, 1067–1068.

Kamiya, T., Tsurumi, T., and Daimon, M. (1993). Quantum calculation of molecular orbitals for PZT solid solutions by DVXα cluster method. In *Computer Aided Innovation of New Materials II, M. Doyama, J. Kihara, M. Tanaka, R. Yamamoto (Editors)*, pages 225–228.

Kanaev, I. F. and Malinovskii, V. K. (1982). Asymmetry of conductivity along the polarization axis in ferroelectrics. *Doklady Akademii Nauk SSSR*, **266**, 1367.

Kanaya, H., Iwamoto, T., Takahagi, Y., Kunishima, I., and Tanaka, S. (1999). Hydrogen induced imprint mechanism of Pt/PZT/Pt capacitors by low-temperature hydrogen treatment. *Integrated Ferroelectrics*, **25**, 575–584.

Kang, B. S., Yoon, Jong-Gul, Noh, T. W., Song, T. K., Seo, S., Lee, Y. K., and Lee, J. K. (2003). Polarization dynamics and retention loss in fatigued $PbZr_{0.4}Ti_{0.6}O_3$ ferroelectric capacitors. *Appl. Phys. Lett.*, **82**, 248–250.

Kang, Dong Heon, Maeng, Yong Joo, Shin, Sang Hyun, Park, Jeong Hwan, and Yoon, Ki Hyun (2001). Crystal structure, microstructure and ferroelectric properties of PZT(55/45) and PZT(80/20) thin films due to various buffer layers. *Ferroelectrics*, **260**, 125–130.

Känzig, W. (1955). Space charge layer near the surface of a ferroelectric. *Phys. Rev.*, **98**, 549–550.

Keve, E. T., Abrahams, S. C., and Bernstein, J. L. (1971). Ferroelectric ferroelastic paramagnetic β-$Gd_2(MoO_4)_3$ crystal structure of the transition-metal molybdates and tungstates. VI. *J. Chem. Phys.*, **54**, 3185–3194.

Khalilov, Sh., Dimza, V., Krumins, A., Sprogis, A., and Spule, A. (1986). Asymmetry of electroconductivity in polarized PLZT ceramics. *Ferroelectrics*, **69**, 59–65.

Khatchaturyan, K. (1995). Mechanical fatigue in thin films induced by piezoelectric strains as a cause of ferroelectric fatigue. *J. Appl. Phys.*, **77**, 6449–6455.

Kholkin, A. L., Colla, E. L., Tagantsev, A. K., Taylor, D. V., and Setter, N. (1996). Fatigue of piezoelectric properties in Pb(Zr,Ti)O_3 films. *Appl. Phys. Lett.*, **68**, 2577–2579.

Kim, Dong-Chun and Lee, Won-Jong (2002). Effect of $LaNiO_3$ top electrode on the resistance of Pb(Zr,Ti)O_3 ferroelectric capacitor to hydrogen damage and fatigue. *Jpn. J. Appl. Phys.*, **41**, 1470–1476.

Kim, Hong Koo and Basit, N. A. (2000). Ferroelectric nonvolatile memory field-effect transistors based on a novel buffer layer structure. *Int. J. High Speed Electron. Syst.*, **10**, 39–46.

Kim, Ill Won, Bae, Sang Bo, and An, Chang Won (2001a). Temperature effects on fatigue and retention behaviors of $SrBi_2Ta_2O_9$ thin films fabricated by pulsed laser deposition. *Ferroelectrics*, **260**, 261–266.

Kim, Joon Ham and Park, Chiang Yub (1995). Ferroelectric fatigue characteristics of yttrium added $Pb(Zr_{0.65}Ti_{0.35})O_3$ thin films prepared by sol-gel processing. *Ferroelectrics*, **173**, 37–43.

Kim, S. H., Hong, J. G., Streiffer, S. K., and Kingon, A. I. (1999a). The effect of RuO_2/Pt hybrid bottom electrode structure on the leakage and fatigue properties of chemical solution derived $Pb(Zr_xTi_{1-x})O_3$ thin films. *J. Mater. Res.*, **14**, 1018–1025.

Kim, S. S., Kang, T. S., and Je, J. H. (2000a). Structure and properties of (001)-oriented $Pb(Zr,Ti)O_3$ films on $LaNiO_3$/Si(001) substrates by pulsed laser deposition. *J. Mater. Res.*, **15**, 2881–2886.

Kim, Seung-Hyun, Kim, Dong-Joo, Hong, Joon Goo, Streiffer, S. K., and Kingon, A. I. (1999b). Imprint and fatigue properties of chemical solution derived $Pb_{1-x}La_x(Zr_yTi_{1-y})_{1-x/4}O_3$ thin films. *J. Mater. Res.*, **14**, 1371–1377.

Kim, Sung Jin, Kang, Seong Jun, and Yoon, Yung Sup (2001b). Effect of Zr/Ti concentration in the PLZT(10/y/z) thin films from the aspect of NVFRAM application. *J. Inst. Electr. Eng. Korea SD*, **38**, 1–10.

Kim, Woo Sik, Kim, Ji-Wan, Park, Hyung-Ho, and Lee, Ho Nyung (2000b). Fabrication and characterization of Pt-oxide electrode for ferroelectric random access memory application. *Jpn. J. Appl. Phys.*, **39**, 7097–7099.

Kimura, S., Izumi, K., and Tatsumi, T. (2002). Tetragonal distortion of c domains in fatigued $Pb(Zr,Ti)O_3$ thin films determined by x-ray diffraction measurements with highly brilliant synchrotron radiation. *Appl. Phys. Lett.*, **80**, 2365–2367.

Kingery, W. D., Bowen, H. K., and Uhlmann, D. R. (1976). *Introduction to Ceramics.* John Wiley & Sons, New York.

Kirchhoff, G., Pompe, W., and Bahr, H. A. (1982). Sructure dependence of thermally induced microcracking in porcelain studied by acoustic emission. *J. Mater. Sci.*, **17**, 2809–2816.

Klee, M., De Veirman, A., and Mackens, U. (1993). Analytical study of the growth of polycrystalline titanate thin films. *Philips J. Res.*, **47**, 263.

Klie, R. F. and Browning, N. D. (2000). Atomic scale characterization of oxygen vacancy segregation at $SrTiO_3$ grain boundaries. *Appl. Phys. Lett.*, **77**, 3737–3739.

Kobayashi, J., Someya, T., and Furuhata, Y. (1972). Application of electron-mirror microscopy to direct observation of moving ferroelectric domains of $Gd_2(MoO_4)_3$. *Phys. Lett.*, **38A**, 309–310.

Kolmogorov, A. N. (1937). *Izv. Akad. Nauk. Ser. Math., Bulletin of the Academy of Sciences of the USSR- Mathematical Series*, **3**, 355.

Koo, Sang-Mo, Zheng, Li-Rong, and Rao, R. V. (1999). $BaRuO_3$ thin film electrode for ferroelectric lead zirconate titanate capacitors. *J. Mater. Res.*, **14**, 3833–3836.

Krell, A. and Kirchhoff, G. (1985). Acoustic emission on macroscopic fracture of different Al_2O_3 structures. *J. Mater. Sci. Lett.*, **4**, 1524–1526.

Krishnan, A., Treacy, M. M. J., Bisher, M. E., Chandra, P., and Littlewood, P. B. (2000a). Displacement charge patterns and ferroelectric domain wall dynamics by in-situ TEM. In *Ferroelectric Thin Films VIII, R.W. Schwartz, P.C. McIntyre, Y. Miyasaka, S.R. Summerfelt, D. Wouters, eds., Mater. Res. Soc. Symp. Proc.*, volume 596, pages 161–166.

Krishnan, A., Treacy, M. M. J., Bisher, M. E., Chandra, P., and Littlewood, P. B. (2000b). Maxwellian charge on domain walls. In *Fund. Phys. Ferroelectrics 2000: Aspen Center Phys. Wkshp., R.E. Cohen, Ed., Am. Inst. Phys. Conf. Proc.*, volume 535, pages 191–200.

Krishnan, A., Treacy, M. M. J., Bisher, M. E., Chandra, P., and Littlewood, P. B. (2002). Efficient switching and domain interlocking observed in polyaxial ferroelectrics. *Integrated Ferroelectrics*, **43**, 31–49.

Kröger, F. A. and Vink, H. J. (1956). Relations between the concentrations of imperfections in crystalline solids. In *Solid State Physics, Eds. E. Seitz and D. Turnbull, Academic Press, NY*, volume 3, pages 307–435.

Kroupa, F., Nejezchleb, K., and Saxl, I. (1988). Anisotropy of internal stresses in poled PZT ceramics. *Ferroelectrics*, **88**, 123–137.

Kroupa, F., Nejezchleb, K., Rataj, J., and Saxl, I. (1989). Non-homogeneous distribution of internal stresses and cracks in poled PZT ceramics. *Ferroelectrics*, **100**, 281–290.

Kudzin, A. Yu. and Panchenko, T. V. (1972). Stabilization of spontaneous polarization of $BaTiO_3$ single crystal. *Sov. Phys. Solid State*, **14**, 1599–1600.

Kudzin, A. Yu., Panchenko, T. V., and Yudin, S. P. (1975). Behavior of 180° domain walls of barium titanate single crystal during the "fatigue" and recovery of switching polarization. *Sov. Phys. Solid State*, **16**, 1589–1590.

Kumada, A. (1969). Domain switching in $Gd_2(MoO_4)_3$. *Phys. Lett.*, **30A**, 186–187.

Kundu, T. K. and Lee, Joseph Ya-Min (2000). Thickness-dependent electrical properties of $Pb(Zr,Ti)O_3$ thin film capacitors for memory device applications. *J. Electrochem. Soc.*, **147**, 326–329.

Lambeck, P. V. and Jonker, G. H. (1978). Ferroelectric domain stabilization in $BaTiO_3$ by bulk ordering of defects. *Ferroelectrics*, **22**, 729–731.

Lambeck, P. V. and Jonker, G. H. (1986). The nature of domain stabilization in ferroelectric perowskites. *J. Phys. Chem. Solids*, **47**, 453–461.

Lawn, B. (1993). *Fracture of brittle solids.* Cambridge University Press, Cambridge, UK.

Lee, J., Johnson, L., Safari, A., Ramesh, R., Sands, T., Gilchrist, H., and Keramidas, V. G. (1993). Effect of crystalline quality and electrode material on fatigue in $Pb(Zr,Ti)O_3$ thin film capacitors. *Appl. Phys. Lett.*, **63**, 27–29.

Lee, J., Esayan, S., Safari, A., and Ramesh, R. (1995a). Fatigue and photoresponse of lead zirconate titanate thin film capacitors. *Integrated Ferroelectrics*, **6**, 289–300.

Lee, J. J., Thio, C. L., and Desu, S. B. (1995b). Electrode contacts on ferroelectric $Pb(Zr_{(x)}Ti_{(1-x)})O_3$ and $SrBi_2Ta_2O_9$ thin films and their influence on fatigue properties. *J. Appl. Phys.*, **78**, 5073–5077.

Lee, J. K., Hong, K. S., and Chung, J. H. (2001a). Revisit to the origin of grain growth anomaly in yttria-doped barium titanate. *J. Am. Ceram. Soc.*, **84**, 1745–1749.

Lee, Jang-Sik and Joo, Seung-Ki (2002). Analysis of grain-boundary effects on the electrical properties of $Pb(Zr,Ti)O_3$ thin films. *Appl. Phys. Lett.*, **81**, 2602–2604.

Lee, Jang-Sik, Park, Eung-Chul, Park, Jung-Ho, Lee, Byung-Il, and Joo, Seung-Ki (2000a). Effects of grain boundary on the ferroelectric properties of selectively grown PZT thin films. In *Ferroelectric Thin Films VIII, R.W. Schwartz, P.C. McIntyre, Y. Miyasaka, S.R. Summerfelt, D. Wouters, eds., Mater. Res. Soc. Symp. Proc.*, volume 596, pages 315–320.

Lee, Jang-Sik, Kim, Chan-Soo, and Joo, Seung-Ki (2000b). Fatigue and data retention characteristics of single-grained $Pb(Zr,Ti)O_3$ thin films. In *IEEE Proc. 12th Intern. Symp. on the Appl. of Ferroelectrics, ISAF*, pages 595–598.

Lee, Jang-Sik, Park, Eung-Chul, Park, Jung-Ho, Lee, Byung-Il, and Joo, Seung-Ki (2000c). A study on the selective nucleations for formation of large single grains in PZT thin films. In *Ferroelectric Thin Films VIII, R.W. Schwartz, P.C. McIntyre, Y. Miyasaka, S.R. Summerfelt, D. Wouters, eds., Mater. Res. Soc. Symp. Proc.*, volume 596, pages 217–222.

Lee, Jang-Sik, Kim, Chan-Soo, and Joo, Seung-Ki (2001b). Fatigue and data retention characteristics of single-grained $Pb(Zr,Ti)O_3$ thin films. *J. Korean Phys. Soc.*, **39**, 184–188.

Lee, Su-Jae, Moon, Seung-Eon, Kim, Won-Jeong, and Kim, Eun-Kyong (2001c). Ultralow-frequency dielectric relaxation and fatigue properties of PZT thin film: Effect of space charges. *Integrated Ferroelectrics*, **37**, 225–234.

Li, X., Vartuli, J. S., Milius, D. L., Aksay, I. A., Shih, W. Y., and Shih, W. H. (2001). Electromechanical properties of a ceramic d31-gradient flextensional actuator. *J. Am. Ceram. Soc.*, **84**, 996–1003.

Lin, Wen-Jen, Tseng, Tseng-Yuen, Lin, Shuen-Perg, Tu, Shun-Lih, Chang, Hong, Yang, Sheng-Jenn, and Lin, I-Nan (1997). Influence of crystal structure on the fatigue properties of $Pb_{1-x}La_x(Zr_yTi_z)O_3$ thin films prepared by pulsed-laser deposition technique. *J. Am. Ceram. Soc.*, **80**, 1065–1072.

Lines, M. E. and Glass, A. M. (1977). *Principles and Applications of Ferroelectrics and related Materials.* Clarendon Press, Oxford.

Liu, Dage, Wang, Chen, Zhang, Hongxi, Li, Junwei, Zhao, Liancheng, and Bai, Chunli (2001). Domain configuration and interface structure analysis of sol-gel-derived PZT ferroelectric thin films. *Surf. Interf. Anal.*, **32**, 27–31.

Lo, V. C. and Chen, Z. J. (2001a). Modelling the roles of oxygen vacancies in thin film ferroelectric memory. In *ISAF 2000. Proceedings of the 12th IEEE Int. Symp. on Appl. of Ferroelectrics 2000 (IEEE Cat. No. 00CH37076)*, pages 157–160.

Lo, Veng Cheong and Chen, Zhijiang (2001b). Investigation of the role of oxygen vacancies on polarization fatigue in ferroelectric thin films. *Ferroelectrics*, **259**, 145–150.

Lohkämper, R., Neumann, H., and Arlt, G. (1990). Internal bias in acceptor-doped $BaTiO_3$ ceramics: Numerical evaluation of increase and decrease. *J. Appl. Phys.*, **68**, 4220–4224.

Lucato, S. L. dos e. Santos, Lupascu, D. C., Kamlah, M., Rödel, J., and Lynch, C. S. (2001a). Constraint induced crack initiation at electrode edges. *Acta Mater.*, **49**, 2751–2759.

Lucato, S. L. S., Lupascu, D. C., and Rödel, J. (2001b). Crack initiation and crack propagation in partially electroded PZT. *J. Eur. Ceram. Soc.*, **21**, 1424–1429.

Lupascu, D. C. (2001). Microcracking and discontinuous fast switching as acoustic emission sources in 8/65/35 and 9.5/65/35 PLZT relaxor ferroelectrics. *J. Eur. Ceram. Soc.*, **21**, 1429–1432.

Lupascu, D. C. and Hammer, M. (2002). Discontinuous switching, dynamic relaxation, and microcracking in lead-zirconate-titanate monitored by acoustic emissions. *phys. stat. sol. (a)*, **191**, 643–657.

Lupascu, D. C. and Rabe, U. (2002). Cyclic cluster growth in ferroelectric perovskites. *Phys. Rev. Lett.*, **89**, 187601.

Lupascu, D. C., Nuffer, J., and Rödel, J. (2000). Role of crack formation in the electric fatigue behavior of ferroelectric PZT ceramics. In *SPIE Smart Struct.*

and Mater. 2000: Active Materials: Behavior and Mechanics, C. S. Lynch, Ed., volume 3992, pages 209–216.

Lupascu, D. C., Nuffer, J., and Rödel, J. (2001). Microcrack fatigue trees in 9.5/65/35 PLZT. *J. Europ. Ceram. Soc.*, **21**, 1421–1423.

Lupascu, D. C., Fedosov, S., Verdier, C., v. Seggern, H., and Rödel, J. (2004). Stretched exponential relaxation in perovskite ferroelectrics after cyclic loading. *J. Appl. Phys., subm.*

Maier, J. (1996). Chemical diffusion of oxygen in perovskites. *Phase Transitions*, **58**, 217–234.

Majumder, S. B., Agrawal, D. C., Mohapatra, Y. N., and Katiyar, R. S. (2000a). Fatigue and dielectric properties of undoped and Ce doped PZT thin films. *Integrated Ferroelectrics*, **29**, 63–74.

Majumder, S. B., Bhaskar, S., Dobal, P. S., and Katiyar, R. S. (2000b). Investigation on the electrical characteristics of sol-gel derived $Pb_{1.05}(Zr_{0.35}Ti_{0.47})O_3$ thin films. *Integrated Ferroelectrics*, **29**, 87–98.

Majumder, S. B., Roy, B., Katiyar, R. S., and Krupanidhi, S. B. (2001). Improvement of the degradation characteristics of sol-gel derived PZT (53/47) thin films: Effect of conventional and graded iron doping. *Integrated Ferroelectrics*, **39**, 127–136.

Malén, K. and Bolin, L. (1974). A theoretical estimate of acoustic-emission stress amplitudes. *phys. stat. sol. (b)*, **61**, 637–645.

Malvern, L. E. (1969). *Introduction to the Mechanics of a Continuous Medium.* Prentice-Hall, Englewood Cliffs, NJ, USA.

Mansour, S. A. and Vest, R. W. (1992). The dependence of ferroelectric and fatigue behaviours of PZT films on microstructure and orientation. *Integrated Ferroelectrics*, **1**, 57–69.

McQuarrie, M. (1953). Time effects in the hysteresis loop of polycrystalline barium titanate. *J. Appl. Phys.*, **24**, 1334–1335.

McQuarrie, M. C. and Buessem, W. R. (1955). The aging effect in barium titanate. *Ceramic Bulletin*, **34**, 402–406.

Mecartney, M. L., Sinclair, R., and Ewell, G. J. (1980). Chemical and microstructural analyses of grain boundaries in $BaTiO_3$-based dielectrics. *Adv. Ceram.*, **1**, 207–214.

Mehta, R. R., Silverman, B. D., and Jacobs, J. T. (1973). Depolarization fields in thin ferroelectric films. *J. Appl. Phys.*, **44**, 3379–3385.

Meng, X. J., Sun, J. L., Yu, J., Wang, G. S., Guo, S. L., and Chu, J. H. (2001). Enhanced fatigue property of PZT thin films using $LaNiO_3$ thin layer as bottom electrode. *Appl. Phys. A*, **73**, 323–325.

Merz, W. J. and Anderson, J. R. (1955). Ferroelectric storage device. *Bell Lab. Rec.*, **33**, 335–342.

Mihara, T., Watanabe, H., and Araujo, C. A. Paz de (1994a). Characteristic change due to polarization fatigue of sol-gel ferroelectric $Pb(Zr_{0.4}Ti_{0.6})O_3$ thin-film capacitors. *Jpn. J. Appl. Phys.*, **33**, 5281–5286.

Mihara, T., Watanabe, H., and Araujo, C. A. Paz de (1994b). Polarization fatigue characteristics of sol-gel ferroelectric $Pb(Zr_{0.4}Ti_{0.6})O_3$ thin-film capacitors. *Jpn. J. Appl. Phys.*, **33, part 1, 7A**, 3996–4002.

Miller, D. W. and Glower, D. D. (1972). The properties of oxygen depleted ceramic $Pb(Zr_{0.65}Ti_{0.35})O_3$+1wt% Nb_2O_5 surface layers. *Ferroelectrics*, **3**, 295–303.

Miller, R. C. and Weinreich, G. (1960). Mechanism for the sidewise motion of 180° domain walls in barium titanate. *Phys. Rev.*, **117**, 1460–1466.

Mitsui, T., Tatsuzaki, I., and Nakamura, E. (1976). *An Introduction to the Physics of Ferroelectrics*. Gordon and Breach Publishers, New York.

Miura, K. and Tanaka, M. (1996a). Effect of La-doping on fatigue in ferroelectric perovskite oxides. *Jpn. J. Appl. Phys.*, **35**, 3488–3491.

Miura, K. and Tanaka, M. (1996b). Origin of fatigue in ferroelectric perovskite oxides. *Jpn. J. Appl. Phys.*, **35**, 2719–2725.

Mohamad, I. J., Lambson, E. F., Miller, A. J., and Saunders, G. A. (1979). A threshold electric field for acoustic emission from ferroelectric $Pb_5Ge_5O_{11}$. *Phys. Lett.*, **71 A**, 115.

Mohamad, I. J., Zammit-Mangion, L. J., Lambson, E. F., Saunders, G. A., and Abey, A. E. (1981). Acoustic emission from ferroelectric crystals. *Ferroelectrics*, **38**, 983.

Mohamad, I. J., Mangion, L. Z., Lambson, E. F., and Saunders, G. A. (1982). Acoustic emission from domain wall motion in ferroelectric lead germanate. *J. Phys. Chem. Solids*, **43**, 749–759.

Morosova, G. P. and Serdobolskaja, O. Y. (1987). Acoustical emission by switching gadolinium molybdate. *Ferroelectrics*, **75**, 449–453.

Morse, P. M. and Ingard, K. U. (1968). *Theoretical Acoustics*. McGraw-Hill Book Company, NY.

Nagaraj, B., Aggarwal, S., and Ramesh, R. (2001). Influence of contact electrodes on leakage characteristics in ferroelectric thin films. *J. Appl. Phys.*, **90**, 375–382.

Nagata, H., Haneda, H., Sakaguchi, I., Takenaka, T., and Tanaka, J. (1997). Reaction and diffusion between PLZT ceramics and Ag electrode. *J. Ceram. Soc. Jpn.*, **105**, 862–867.

Nagata, H., Uemastu, H., Sakaguchi, I., Haneda, H., and Takenaka, T. (2001). Fatigue properties of bismuth titanate annealed in the reduced atmosphere. *Ferroelectrics*, **257**, 203–210.

Nejezchleb, K., Kroupa, F., Boudys, M., and Zelenka, J. (1988). Poling process, microcracks, and mechanical quality of PZT ceramics. *Ferroelectrics*, **81**, 339–342.

Neumann, H. and Arlt, G. (1986a). Maxwell-Wagner relaxation and degradation of $SrTiO_3$ and $BaTiO_3$ ceramics. *Ferroelectrics*, **69**, 179–86.

Neumann, H. and Arlt, G. (1986b). Transient polarization and degradation effects in $BaTiO_3$ and $SrTiO_3$ ceramics. In *Proc. IEEE Int. Symp. on Appl. of Ferroelectrics*, pages 357–360.

Newnham, R. E., McKinstry, H. A., Gregg, C. W., and Stitt, W. R. (1969). Lattice parameters of ferroelectric rare earth molybdates. *phys. stat. sol.*, **32**, K49–K51.

Noh, T. W., Park, B. H., Kang, B. S., Bu, S. D., and Lee, J. (2000). A new ferroelectric material for use in FRAM: Lanthanum-substituted bismuth titanate. In *IEEE Proc. 12th Intern. Symp. on the Appl. of Ferroelectrics, ISAF2000*, pages 237–242.

Nozawa, Hiroshi, Takayama, Masao, and Koyama, Shinzo (2001). Ferroelectric dielectric technology. In *2001 6th Int. Conf. Solid-State Integr. Circuit Techn. Proc. IEEE*, pages 687–691.

Nuffer, J., Lupascu, D. C., and Rödel, J. (2000). Damage evolution in ferroelectric PZT induced by bipolar electric cycling. *Acta Mater.*, **48**, 3783–3794.

Nuffer, J., Schröder, M., Lupascu, D. C., and Rödel, J. (2001). Negligible oxygen liberation during the fatigue of lead-zirconate-titanate. *Appl. Phys. Lett.*, **79(22)**, 3675–3677.

Ohtsu, M. (1995). Acoustic emission theory for moment tensor analysis. *Res. Nondestr. Eval.*, **6**, 169–184.

Ozgul, M., Takemura, K., Trolier-McKinstry, S., and Randall, C. A. (2001). Polarization fatigue in $Pb(Zn_{1/3}Nb_{2/3})O_3$-$PbTiO_3$ ferroelectric single crystals. *J. Appl. Phys.*, **89**, 5100–5106.

Palanduz, A. C. and Smyth, D. M. (2000). Defect chemistry of $SrBi_2Ta_2O_9$ and ferroelectric fatigue endurance. *J. Electrocer.*, **5**, 21–30.

Pan, M. J., Park, S. E., Markowski, K. A., Yoshikawa, S., and Randall, C. A. (1996). Superoxidation and electrochemical reactions during switching in $Pb(Zr,Ti)O_3$ ceramics. *J. Am. Ceram. Soc.*, **79**, 2971–2974.

Pan, W., Yue, C. F., and Sun, S. (1992a). Domain orientation change induced by ferroelectric fatigue process in lead zirconate titanate ceramics. *Ferroelectrics*, **133**, 97–102.

Pan, W., Yue, C. F., and Tosyali, O. (1992b). Fatigue of ferroelectric polarization and the electric field induced strain in lead lanthanum zirconate titanate ceramics. *J. Am. Ceram. Soc.*, **75**, 1534–1540.

Pan, W. Y., Dam, C. Q., Zhang, Q. M., and Cross, L. E. (1989). Large displacement transducers based on electric field forced phase transitions in the tetragonal $(Pb_{0.97}La_{0.02})(Ti,Zr,Sn)O_3$ family of ceramics. *J. Appl. Phys.*, **66**, 6014–6023.

Pan, W. Y., Yue, C. F., and Tuttle, B. A. (1992c). Ferroelectric fatigue in modified bulk lead zirconate titanate ceramics and thin films. In *Ceramic Transactions, Ferroelectric Films, A.S. Bhalla, K.M. Nair, Eds.*, volume 25, pages 385–397.

Park, B. H., Kang, B. S., Bu, S. D., Noh, T. W., Lee, J., and Jo, W. (1999). Lanthanum-substituted bismuth titanate for use in non-volatile memories. *Nature*, **401**, 682.

Park, C. H. and Chadi, D. J. (1998). Microscopic study of oxygen-vacancy defects in ferroelectric perowskites. *Phys. Rev. B*, **57**, R13961–R13964.

Paton, E., Brazier, M., Mansour, S., and Bement, A. (1997). A critical study of defect migration and ferroelectric fatigue in lead zirconate titanate thin film capacitors under extreme temperatures. *Integrated Ferroelectrics*, **18**, 29–37.

Paulik, S. W., Zimmermann, M. H., and Faber, K. T. (1996). Residual stress in ceramics with large thermal expansion anisotropy. *J. Mater. Res.*, **11**, 2795–2803.

Pawlaczyk, CZ., Tagantsev, A. K., Brooks, K., Reaney, I. M., Klissurska, R., and Setter, N. (1995). Fatigue, rejuvenation and self-restoring in ferroelectric thin films. *Integrated Ferroelectrics*, **8**, 293–316.

Peterson, C. R., Mansour, S. A., and Bement, Jr., A. (1995). Effects of optical illumination on fatigued lead zirconate titanate capacitors. *Integrated Ferroelectrics*, **7**, 139–147.

Pferner, R. A., Thurn, G., and Aldinger, F. (1999). Mechanical properties of PZT ceramics with tailored microstructure. *Mater. Chem. Phys.*, **61**, 24–30.

Pike, G. E., Warren, W. L., Dimos, D., Tuttle, B. A., Ramesh, R., Lee, J., Keramides, V. G., and Evans, Jr., J. T. (1995). Voltage offsets in $(Pb,La)(Zr,Ti)O_3$ thin films. *Appl. Phys. Lett.*, **66**, 484–486.

Plessner, K. W. (1956). Aging of the dielectric properties of barium titanate ceramics. *Proc. Phys. Soc. B*, **69**, 1261–1268.

Polcawich, R. G. and Trolier-McKinstry, S. (2000). Piezoelectric and dielectric reliability of lead zirconate titanate thin films. *J. Mater. Res.*, **15(11)**, 2505–2513.

Pöykkö, S. and Chadi, D. J. (2000). First principles study of Pb vacancies in $PbTiO_3$. *Appl. Phys. Lett.*, **76**, 499–501.

Prisedsky, V. V., Shishkovsky, V. I., and Klimov, V. V. (1978). High-temperature electrical conductivity and point defects in lead zirconate-titanate. *Ferroelectrics*, **17**, 465–468.

Prokopalo, O. I. (1979). Point defects, electrical conductivity, and electron energy spectra of perovskite oxides. *Sov. Phys. Solid State*, **21(10)**, 1768–1770.

Ramesh, R. and Keramidas, V. G. (1995). Metal-oxide heterostructures. *Annu. Rev. Mater. Sci.*, **25**, 647–678.

Ramesh, R., Chan, W. K., Gilchrist, H., Wilkens, B., Sands, T., Tarascon, J. M., Keramidas, V. G., Evans, Jr., J. T., Gealy, F. Dan, and Fork, D. K. (1992). Oxide ferroelectric/cuprate superconductor heterostructures: Growth and properties. *Mater. Res. Soc. Symp. Proc.*, **243**, 477–487.

Raymond, M. V. and Smyth, D. M. (1993). Defects and transport in $Pb(Zr_{1/2}Ti_{1/2})O_3$. *Ferroelectrics*, **144**, 129–135.

Raymond, M. V. and Smyth, D. M. (1994). Defect chemistry and transport properties of $Pb(Zr_{1/2}Ti_{1/2})O_3$. *Integrated Ferroelectrics*, **4**, 145–154.

Reed-Hill, R. E. and Abbaschian, R. (1994). *Physical Metallurgy Principles.* PWS Publishing Company, Boston, USA.

Ren, Tian-Ling, Zhang, Lin-Tao, Liu, Li-Tian, and Li, Zhi-Jian (2002). Preparation and characterization of a sol-gel prepared ferroelectric sandwich structure. *J. Sol-Gel Sci. Tech.*, **24**, 271–274.

Ricinschi, D., Lerescu, A. I., and Okuyama, M. (2000). Investigation of fatigue mechanisms in $Pb(Zr,Ti)O_3$ films from a correlated analysis of hysteresis parameters in a lattice model with distributed polarization clamping. *Jpn. J. Appl. Phys.*, **39**, L990–L992.

Robels, U. (1993). *Alterungserscheinungen in ferroelektrischer Keramik.* Verlag Mainz, Aachen.

Robels, U. and Arlt, G. (1993). Domain wall clamping in ferroelectrics by orientation of defects. *J. Appl. Phys.*, **73**, 3454–3460.

Robels, U., Schneider-Störmann, L., and Arlt, G. (1995a). Dielectric aging and its temperature dependence in ferroelectric ceramics. *Ferroelectrics*, **168**, 301–311.

Robels, U., Calderwood, J. H., and Arlt, G. (1995b). Shift and deformation of the hysteresis curve of ferroelectrics by defects: An electrostatic model. *J. Appl. Phys.*, **77**, 4002.

Robertson, J. and Warren, W. L. (1995). Energy levels of point defects in perovskite oxides. *Mater. Res. Soc. Proc.*, **361**, 123–128.

Robertson, J., Warren, W. L., Tuttle, B. A., Dimos, D., and Smyth, D. M. (1993). Shallow Pb^{3+} hole traps in lead zirconate titanate ferroelectrics. *Appl. Phys. Lett.*, **63**, 1519–1521.

Rouby, D. and Fleischmann, P. (1978). Spectral analysis of acoustic emission from aluminium single crystals undergoing plastic deformation. *phys. stat. sol. (a)*, **48**, 439–445.

Saito, Y. and Hori, S. (1994). Acoustic emission and domain switching in tetragonal lead zirconate titanate ceramics. *Jpn. J. Appl. Phys.*, **33**, 5555–5558.

Salaneck, W. R. (1972). Some fatiguing effects in 8/65/35 PLZT fine grained ferroelectric ceramic. *Ferroelectrics*, **4**, 97–101.

Salje, E. K. H. (1990). *Phase transitions in ferroelastic and co-elastic crystals.* Cambridge University Press, Cambridge.

Sandomirskii, V. B., Khalilov, Sh. S., and Chensky, E. V. (1982). The anomalous photovoltage in a model of the highly doped and compensated ferroelectric semiconductors. *Ferroelectrics*, **43**, 147–151.

Schneider, G. A., Rostek, A., Zickgraf, B., and Aldinger, F. (1994). Crack growth in ferroelectric ceramics under mechanical and electrical loading. In *Electroceramics IV, Vol II, R. Waser et al. Eds., Augustinus Buchhandlung, Aachen,* pages 1211–1216.

Scott, J. F. (2000). *Ferroelectric Memories.* Springer, Berlin, Heidelberg.

Scott, J. F. (2001). Fatigue as a phase transition. *Integrated Ferroelectrics*, **38**, 125–133.

Scott, J. F. and Araujo, C. A. Paz de (1989). Ferroelectric memories. *Science*, **246**, 1400–1405.

Scott, J. F. and Dawber, M. (2000a). Atomic-scale and nanoscale self-patterning in ferroelectric thin films. In *Fund. Phys. Ferroelectrics 2000: Aspen Center Phys. Wkshp., R.E. Cohen, Ed., Am. Inst. Phys. Conf. Proc.*, volume 535, pages 129–135.

Scott, J. F. and Dawber, M. (2000b). Oxygen-vacancy ordering as a fatigue mechanism in perovskite ferroelectrics. *Appl. Phys. Lett.*, **76**, 3801–3803.

Scott, J. F. and Dawber, M. (2002). Physics of ferroelectric thin-film memory devices. *Ferroelectrics*, **265**, 119–128.

Scott, J. F., Araujo, C. A., Brett-Meadow, H., McMillan, L. D., and Shawabkeh, A. (1989). Radiation effects on ferroelectric thin-film memories: Retention failure mechanisms. *J. Appl. Phys.*, **66**, 1444–1453.

Scott, J. F., Araujo, C. A. Paz de, Melnick, B. M., McMillan, L. D., and Zuleeg, R. (1991). Quantitative measurement of space- charge effects in lead zirconate-titanate memories. *J. Appl. Phys.*, **70**, 382–388.

Scott, J. F., Melnick, B. M., McMillan, L. D., Araujo, C. A. Paz de, and Azuma, M. (1993). Dielectric breakdown in high-ϵ films for ULSI DRAMS. *Ferroelectrics*, **150**, 209–218.

Scott, J. F., Redfern, S. A. T., Zhang, Ming, and Dawber, M. (2001). Polarons, oxygen vacancies, and hydrogen in $Ba_xSr_{1-x}TiO_3$. *J. Eur. Ceram. Soc.*, **21**, 1629–1632.

Scruby, C., Wadley, H., and Sinclair, J. E. (1981). The origin of acoustic emission during deformation of aluminium and an aluminium-magnesium alloy. *Phil. Mag. A*, **44**, 249–274.

Serdobolskaja, O. Yu. and Morozova, G. P. (1998). Sound waves in polydomain ferroelectrics. *Ferroelectrics*, **208-209**, 395–412.

Shannigrahi, S. R. and Jang, Hyun M. (2001). Fatigue-free lead zirconate titanate-based capacitors for nonvolatile memories. *Appl. Phys. Lett.*, **79**, 1051–1053.

Shimada, Y., Inoue, A., Nasu, T., Nagano, Y., Matsuda, A., Arita, K., Uemoto, Y., Fujii, E., and Otsuki, T. (1996). Time dependent leakage current behavior of integrated $Ba_{0.7}Sr_{0.3}TiO_3$ thin film capacitors during stressing. *Jpn. J. Appl. Phys.*, **35**, 4919–24.

Shur, V. Ya., Letuchev, V. V., Popov, Yu. A., and Sarapulov, V. I. (1986). Change in domain structure in lead germanate due to polarization reversal. *Sov. Phys. Crystallogr.*, **30**, 548–550.

Shur, V. Ya., Gruverman, A. L., Kuminov, V. P., and Tonkachova, N. A. (1990). Dynamics of plane domain walls in lead germanate and gadolinium molybdate. *Ferroelectrics*, **111**, 197–206.

Shur, V. Ya., Nikolaeva, E. V., Rumyantsev, E. L., Shishkin, E. I., Subbotin, A. L., and Kozhevnikov, V. L. (1999). Smooth and jump-like dynamics of the plane domain wall in gadolinium molybdate. *Ferroelectrics*, **222**, 323–331.

Shur, V. Ya., Rumyantsev, E. L., Nicolaeva, E. V., Shishkin, E. I., Baturin, I. S., Ozgul, M., and Randall, C. A. (2000). Kintetics of fatigue effect. In *11th Int. Symp. On Integr. Ferroel., Aachen, Mar 12th-15th*, volume acc.

Shur, V. Ya., Nikolaeva, E., Shishkin, E. I., Baturin, I. S., Bolten, D., Lohse, O., and Waser, R. (2001a). Fatigue in PZT thin films. In *Mater. Res. Soc. Symp. Proc.*, volume 655, page CC10.8.1.

Shur, V. Ya., Rumyantsev, E. L., Nikolaeva, E. V., Shishkin, E. I., and Baturin, I. S. (2001b). Kinetic approach to fatigue phenomenon in ferroelectrics. *J. Appl. Phys.*, **90**, 6312–6315.

Shur, V. Ya., Rumyantsev, E. L., Nikolaeva, E. V., Shishkin, E. I., Baturin, I. S., Ozgul, M., and Randall, C. A. (2001c). Kinetics of fatigue effect. *Integrated Ferroelectrics*, **33**, 117–132.

Simmons, J. A. and Wadley, H. N. G. (1984). Theory of acoustic emission from phase transformations. *J. Res. National Bureau Standards*, **89**, 55–64.

Sklarczyk, C. (1992). The acoustic emission analysis of the crack processes in alumina. *J. Eur. Ceram. Soc.*, **9**, 427–435.

Slinkina, M. V., Dontsov, G. I., and Zhukovski, V. M. (1993). Diffusional penetration of silver from electrodes into PZT ceramics. *J. Mater. Sci.*, **28**, 5189.

Smyth, D. M. (1985). Defects and order in perovskite-related oxides. *Ann. Rev. Mater. Sci.*, **15**, 329–357.

Smyth, D. M. (1994). Ionic transport in ferroelectrics. *Ferroelectrics*, **151**, 115–124.

Speich, G. R. and Schwoeble, A. J. (1975). Acoustic emission during phase transformation in steel. *ASTM Spec. Tech. Pub., Am. Soc. Testing Mater.*, **571**, 40–58.

Srikanth, V. and Subbarao, E. C. (1992). Acoustic emission in ferroelectric lead titanate ceramics: Origin and recombination of microcracks. *Acta metall. mater.*, **40**, 1091–1100.

Stauffer, D. (1985). *Introduction to Percolation Theory.* Taylor & Francis, London.

Stewart, W. C. and Cosentino, L. S. (1970). Some optical and electrical switching characteristics of a lead zirconate titanate ferroelectric ceramic. *Ferroelectrics*, **1**, 149–167.

Stolichnov, I. and Tagantsev, A. (1998). Space-charge influenced-injection model for conduction in $Pb(Zr_xTi_{1-x})O_3$ thin films. *J. Appl. Phys.*, **84**, 3216–3225.

Stolichnov, I. and Tagantsev, A. (1999). Physical origin of conduction in PZT thin films. *Ferroelectrics*, **225**, 147–154.

Stolichnov, I., Tagantsev, A., Colla, E. L., and Setter, N. (1999). Tunneling conduction in virgin and fatigued states of PZT films. *Ferroelectrics*, **225**, 125–132.

Stolichnov, I., Tagantsev, A., Colla, E., and Setter, N. (2001a). Charge relaxation at the interfaces of low-voltage ferroelectric film capacitors: Fatigue endurance and size effects. *Ferroelectrics*, **258**, 221–230.

Stolichnov, I., Tagantsev, A., Setter, N., Okhonin, S., Fazan, P., Cross, J. S., Tsukada, M., Bartic, A., and Wouters, D. (2001b). Constant-current study of dielectric breakdown of $Pb(Zr,Ti)O_3$ ferroelectric film capacitors. *Integrated Ferroelectrics*, **32**, 45–54.

Strukov, B. A. and Levanyuk, A. P. (1998). *Ferroelectric Phenomena in Crystals.* Springer, Heidelberg.

Sturman, B. I. (1982). Asymmetry of the electrical conductivity in pyroelectrics. *Sov. Phys. Solid State*, **24**, 1742–1745.

Subbarao, E. C. (1991). Microcracking in ceramics and acoustic emission. *Trans. Indian Ceram. Soc.*, **50**, 109–117.

Sun, Haishan and Chen, Shoutian (2000). Charge-discharge breakdown of dielectric ceramic. In *Proc.* 6th *Int. Conf. Prop. Appl. Diel. Mater., Xi'An Jiaotong Univ. (Cat. No.00CH36347)*, volume 2, pages 1029–1032.

Sun, J. L., Chen, J., Meng, X. J., Yu, J., Bo, L. X., Guo, S. L., and Chu, J. H. (2002). Evolution of rayleigh constant in fatigued lead zirconate titanate capacitors. *Appl. Phys. Lett.*, **80**, 3584–3586.

Suzuki, T., Nishi, Y., and Fujimoto, M. (2000). Ruddlesden-Popper planar faults and nanotwins in heteroepitaxial nonstoichiometric barium titanate thin films. *J. Am. Ceram. Soc.*, **83**, 3185–3195.

Suzuki, T., Ueno, M., Nishi, Y., and Fujimoto, M. (2001). Dislocation loop formation in nonstoichiometric $(Ba,Ca)TiO_3$ and $BaTiO_3$ ceramics. *J. Am. Ceram. Soc.*, **84**, 200–206.

Tagantsev, A. K. and Stolichnov, I. A. (1999). Injection-controlled size effect on switching of ferroelectric thin films. *Appl. Phys. Lett.*, **74**, 1326–1328.

Tagantsev, A. K., Stolichnov, I., Colla, E. L., and Setter, N. (2001). Polarization fatigue in ferroelectric films: Basic experimental findings, phenomenological scenarios, and microscopic features. *J. Appl. Phys.*, **90**, 1387–1402.

Tajiri, M. and Nozawa, H. (2001). New fatigue model based on thermionic field emission mechanism. *Jpn. J. Appl. Phys.*, **40**, 5590–5594.

Takatani, S., Kushida-Abdelghafar, K., and Miki, H. (1997). Effect of H_2 annealing on a Pt/PZT interface studied by x-ray photoelectron spectroscopy. *Jpn. J. Appl. Phys.*, **36**, L435–L438.

Takemura, K., Ozgul, M., Bornard, V., and Trolier-McKinstry, S. (2000). Fatigue anisotropy in single crystal $Pb(Zn_{1/3}Nb_{2/3})O_3$-$PbTiO_3$. *J. Appl. Phys.*, **88**, 7272–7277.

Tan, Xiaoli, Xu, Zhengkui, and Shang, Jian Ku (2001). In situ transmission electron microscopy observations of electric-field-induced switching and microcracking in ferroelectric ceramics. *Mater. Sci. & Engin. A*, **314**, 157–161.

Tanaka, K., Uchiyama, K., Azuma, M., Shimada, Y., Otsuki, T., Joshi, V., and Araujo, C. A. Paz de (2001). MOD preparation and characterization of BLT thin film. *Integrated Ferroelectrics*, **36**, 183–190.

Taylor, D. J., Geerse, J., and Larsen, P. K. (1995). Fatigue of organometallic chemical vapor deposited $PbZr_xTi_{1-x}O_3$ thin films with Ru/RuO_2 and Pt/Pt electrodes. *Thin Solid Films*, **263**, 221–230.

Taylor, G. W. (1967). Electrical properties of niobium-doped ferroelectric Pb(Zr,Sn,Ti)O_3 ceramics. *J. Appl. Phys.*, **38**, 4697–4706.

Taylor, G. W., Keneman, S. A., and Stewart, W. C. (1972). Active compensation for ferroelectric-optical circuits. *Ferroelectrics*, **2**, 101–112.

Teowee, G., Baertlein, C. D., Kneer, E. A., Boulton, J. M., and Uhlmann, D. R. (1995). Effect of top metallization on the fatigue and retention properties of sol-gel PZT thin films. *Integrated Ferroelectrics*, **7**, 149–160.

Thakoor, S. (1994). Enhanced fatigue and retention in ferroelectric thin film memory capacitors by post-top-electrode anneal treatment. *J. Appl. Phys.*, **75**, 5409–5414.

Thakoor, S. and Maserjian, J. (1994). Photoresponse probe of the space charge distribution in ferroelectric lead zirconate titanate thin film memory capacitors. *J. Vac. Sci. Technol. A*, **12**, 295.

Thompson, C., Munkholm, A., Streiffer, S. K., Stephenson, G. B., Ghosh, K., Eastman, J. A., Auciello, O., Bai, G. R., Lee, M. K., and Eom, C. B. (2001). X-ray scattering evidence for the structural nature of fatigue in epitaxial Pb(Zr,Ti)O_3 films. *Appl. Phys. Lett.*, **78**, 3511–3513.

Torii, K., Colla, E., Song, H. W., Tagantsev, A. No, Kwangsoo, and Setter, N. (2001). Size and top electrode-edge effects on fatigue in Pb(Zr,Ti)O_3 capacitors with Pt-electrodes. *Integrated Ferroelectrics*, **32**, 215–224.

Triebwasser, S. (1960). Space charge fields in $BaTiO_3$. *Phys. Rev.*, **118**, 100–105.

Tsuda, N., Nasu, K., Fujimori, A., and Siratori, K. (2000). *Electronic Conduction in Oxides*. Springer-Verlag, Berlin.

Tsukada, M., Cross, J. S., Fujiki, M., Tomotani, M., and Kotaka, Y. (2000). Evaluation of Pb(Zr,Ti)O_3 capacitors with top $SrRuO_3$ electrodes. *Key Engin. Mater.*, **181-182**, 69–72.

Uchino, K. and Aburatani, H. (2000). Field induced acoustic emission in ferroelectric ceramics. In *Electronic Ceramic Materials and Devices, Ceramic Transactions, K. M. Nair, Ed.*, volume 106.

Uchino, Kenji (1997). *Piezoelectric Actuators and Ultrasonic Motors*. Kluwer Academic Publ., Boston, Dordrecht, London.

Vedula, V. R., Glass, S. J., Saylor, D. M., Rohrer, G. S., Carter, W. C., Langer, S. A., and Fuller, Jr., E. R. (2001). Residual stress predictions in polycrystalline alumina. *J. Am. Ceram. Soc.*, **acc.**

Verdier, C., Lupascu, D. C., and Rödel, J. (2002). Stability of defects in lead-zirconate-titanate after unipolar fatigue. *Appl. Phys. Lett.*, **81**, 2596–2599.

Verdier, C., Lupascu, D. C., and Rödel, J. (2003). Unipolar fatigue of ferroelectric lead zirconate titanate. *J. Eur. Ceram. Soc.*, **23**, 1409–1415.

Vijay, D. P. and Desu, S. B. (1993). Electrodes for $PbZr_xTi_{1-x}O_3$ ferroelectric thin films. *J. Electrochem. Soc.*, **140**, 2640–2645.

Vollmann, M. and Waser, R. (1994). Grain boundary defect chemistry of acceptor-doped titanates: Space charge layer width. *J. Am. Ceram. Soc.*, **77**, 235–243.

Vollmann, M. and Waser, R. (1997). Grain- boundary defect chemistry of acceptor-doped titanates: High field effects. *J. Electroceram.*, **1**, 51–64.

Vollmann, M., Hagenbeck, R., and Waser, R. (1997). Grain- boundary defect chemistry of acceptor- doped titanates: Inversion layer and low- field conduction. *J. Am. Ceram. Soc.*, **80**, 2301–2314.

Wadley, H. N. G. and Scruby, C. B. (1981). Acoustic emission source characterization. *Advances in Acoustic Emission, Proc. Int. Conf. On AE, Anaheim, CA, 1979, H.L. Dunegan, W.F. Hartman eds., Dunhart 1982*, pages 125–153.

Wadley, H. N. G. and Scruby, C. B. (1983). Elastic wave radiation from cleavage crack extension. *Int. J. Fracture*, **23**, 111–128.

Wadley, H. N. G. and Simmons, J. A. (1987). Microscopic origins of acoustic emission. In *Acoustic Emission Testing, Ronnie K. Miller, techn. ed.*, pages 64–90.

Wadley, H. N. G., Scruby, C. B., Sinclair, J. E., and Rusbridge, K. (1981). Quantitative acoustic emission for source characterization in metals. *Harwell Reserch Reports, AERE*, **R-10258**, 1–47.

Wang, Donny, Fotinich, Y., and Carman, G. P. (1998). Influence of temperature on the electromechanical and fatigue behaviour of piezoelectric ceramics. *J. Appl. Phys.*, **83**, 5342–5350.

Wang, G. S., Lai, Z. Q., Meng, X. J., Sun, J. L., Yu, J., Guo, S. L., and Chu, J. H. (2001). $PbZr_{0.5}Ti_{0.5}O_3$ thin films prepared on $La_{0.5}Sr_{0.5}CoO_3/LaNiO_3$ heterostructures for integrated ferroelectric devices. In *IEEE 6th Int. Conf. on Solid-State Integr. Circuit Techn.*, pages 718–721.

Wang, Y., Gong, S. X., Jiang, H., and Jiang, Q. (1996). Modeling of domain pinning effect in polycristalline ferroelectric ceramics. *Ferroelectrics*, **182**, 61–68.

Wang, Y., Wong, K. H., and Choy, C. L. (2002). Fatigue problems in ferroelectric thin films. *phys. stat. sol. (a)*, **191**, 482–488.

Warren, W. L. and Dimos, D. (1994). Photoinduced hysteresis changes and charge trapping in $BaTiO_3$ dielectrics. *Appl. Phys. Lett.*, **64**, 866–868.

Warren, W. L., Dimos, D., Tuttle, B. A., and Smyth, D. M. (1994a). Electronic and ionic trapping at domain walls in $BaTiO_3$. *J. Am. Ceram. Soc.*, **77**, 2753–2757.

Warren, W. L., Robertson, J., Dimos, D.B., Tuttle, B.A., and Smyth, D.M. (1994b). Transient hole traps in PZT. *Ferroelectrics*, **153**, 303–308.

Warren, W. L., Dimos, D., Pike, G. E., and Vanheusden, K. (1995a). Alignment of defect dipoles in polycrystalline ferroelectrics. *Appl. Phys. Lett.*, **67**, 1689–1691.

Warren, W. L., Tuttle, B. A., and Dimos, D. (1995b). Ferroelectric fatigue in perovskite oxides. *Appl. Phys. Lett.*, **67**, 1426–1428.

Warren, W. L., Dimos, D., Tuttle, B. A., Pike, G. E., Raymond, M. V., Nasby, R. D., and Evans, Jr., J. T. (1995c). Mechanisms for the suppression of the switchable polarization in PZT and $BaTiO_3$. *Mater. Res. Soc. Symp. Proc.*, **361**, 51–65.

Warren, W. L., Dimos, D., Tuttle, B. A., Pike, G. E., Schwartz, R. W., Clews, P. J., and McIntyre, D. C. (1995d). Polarization suppression in $Pb(Zr,Ti)O_3$ thin films. *J. Appl. Phys.*, **77**, 6695–6701.

Warren, W. L., Dimos, D., Pike, G. E., Tuttle, B. A., Raymond, M. V., Ramesh, R., and Evans, Jr., J. T. (1995e). Voltage shifts and imprint in ferroelectric capacitors. *Appl. Phys. Lett.*, **67**, 866–868.

Warren, W. L., Pike, G. E., Vanheusen, K., Dimos, D., Tuttle, B. A., and Robertson, J. (1996a). Defect-dipole alignment and tetragonal strain in ferroelectrics. *J. Appl. Phys.*, **79**, 9250–9257.

Warren, W. L., Dimos, D., and Waser, R. M. (1996b). Degradation mechanism in ferroelectric and high-permittivity perovskites. *Mater. Res. Soc. Bull.*, **21(7)**, 40–45.

Warren, W. L., Vanheusden, K., Dimos, D., Pike, G. E., and Tuttle, B. A. (1996c). Oxygen vacancy motion in perovskite oxides. *J. Am. Ceram. Soc.*, **79**, 536–538.

Warren, W. L., Robertson, J., Dimos, D., Tuttle, B. A., Pike, G. E., and Payne, D. A. (1996d). Pb displacements in Pb(Zr,Ti)O_3 perovskites. *Phys. Rev. B*, **53**, 3080–3087.

Warren, W. L., Tuttle, B. A., Rong, F. C., Gerardi, G. J., and Poindexter, E. H. (1997a). Electron paramagnetic resonance investigation of acceptor centers in Pb(Zr,Ti)O_3 ceramics. *J. Am. Ceram. Soc.*, **80**, 680–684.

Warren, W. L., Pike, G. E., Tuttle, B. A., and Dimos, D. (1997b). Polarization-induced trapped charge in ferroelectrics. *Appl. Phys. Lett.*, **70**, 2010–2012.

Warren, W. L., Dimos, D., Tuttle, B. A., Pike, G. E., and Al-Shareef, H. N. (1997c). Relationships among ferroelectric fatigue, electronic charge trapping, defect-dipoles, and oxygen vacancies in perovskite oxides. *Integrated Ferroelectrics*, **16**, 77–86.

Waser, R. (1991). Bulk conductivity and defect chemistry of acceptor-doped strontium titanate in the quenched state. *J. Am. Ceram. Soc.*, **74**, 1934–1940.

Waser, R. (1994). Charge transport in perovskite-type titanates: Space charge effects in ceramics and films. *Ferroelectrics*, **151**, 125–131.

Waser, R. and Hagenbeck, R. (2000). Grain boundaries in dielectric and mixed-conducting ceramics. *Acta Mater.*, **48**, 797–825.

Waser, R. and Smyth, D. M. (1996). Defect chemistry, conduction and breakdown mechanism of perovskite-structure titanates. In *Ferroelectric Thin Films: Synthesis and Basic Properties, C. Paz de Araujo, J.F. Scott, G.W. Taylor, Eds., Ferroelectricity and Related Phenomena, Vol. 10, Gordon and Breach, Amsterdam*, pages 47–92.

Waser, R., Baiatu, T., and Härdtl, K. H. (1990a). d.c. electrical degradation of perovskite-type titanates: I, ceramics. *J. Am. Ceram. Soc.*, **73**, 1645–1653.

Waser, R., Baiatu, T., and Härdtl, K. H. (1990b). d.c. electrical degradation of perovskite-type titanates: II, single crystals. *J. Am. Ceram. Soc.*, **73**, 1654–1662.

Weitzing, H., Schneider, G. A., Steffens, J., Hammer, M., and Hoffmann, M. J. (1999). Cyclic fatigue due to electric loading in ferroelectric ceramics. *J. Eur. Ceram. Soc.*, **19**, 1333–1337.

Wernicke, R. (1976). Part IV. The kinetics of equilibrium restoration in barium titanate. *Philips Res. Repts.*, **31**, 526–543.

Wersing, W. (1974). Analysis of phase mixtures in ferroelectric ceramic by dielectric measurements. *Ferroelectrics*, **7**, 163–165.

White, G. (2000). private communication.

Williams, R. (1965). Surface layer and decay of the switching properties of barium titanate. *J. Phys. Chem. Solids*, **26**, 399–405.

Winzer, S. R., Shankar, N., and Ritter, A. P. (1989). Designing cofired multilayer electrostrictive actuators for reliability. *J. Am. Ceram. Soc.*, **72**, 2246–2257.

Wu, Di, Li, Aidong, Ling, Huiqin, Yu, Tao, Liu, Zhiguo, and Ming, Naiben (2000a). Fatigue study of metalorganic-decomposition-derived $SrBi_2Ta_2O_9$ thin films: The effect of partial switching. *Appl. Phys. Lett.*, **76**, 2208–2210.

Wu, Di, Li, Aidong, Zhu, Tao, Li, Zhifeng, Liu, Zhiguo, and Ming, Naiben (2001). Processing- and composition dependent characteristics of chemical solution deposited $Bi_{4-x}La_xTi_3O_{12}$ thin films. *J. Mater. Res.*, **16**, 1325–1332.

Wu, K. and Schulze, W. A. (1992a). Aging of the weak-field dielectric response in fine- and coarse-grain ceramic $BaTiO_3$. *J. Am. Ceram. Soc.*, **75**, 3390–95.

Wu, K. and Schulze, W. A. (1992b). Effect of the ac field level on the aging of the dielectric response in polycristalline $BaTiO_3$. *J. Am. Ceram. Soc.*, **75**, 3385–89.

Wu, Wenbin, Wong, K. H., Mak, C. L., Choy, C. L., and Zhang, Y. H. (2000b). Effect of oxygen stoichiometry on the ferroelectric property of epitaxial all-oxide $La_{0.7}Sr_{0.3}MnO_3/Pb(Zr_{0.52}Ti_{0.48})O_3/La_{0.7}Sr_{0.3}MnO_3$ thin-film capacitors. *J. Vac. Sci. Technol. A*, **18**, 2412–2416.

Wu, Wenbin, Wong, K. H., Choy, C. L., and Zhang, Y. H. (2000c). Top-interface-controlled fatigue of epitaxial $Pb(Zr_{0.52}Ti_{0.48})O_3$ ferroelectric thin films on $La_{0.7}Sr_{0.3}MnO_3$ electrodes. *Appl. Phys. Lett.*, **77**, 3441–3443.

Wu, Z. and Sayer, M. (1992). Defect structures and fatigue in ferroelectric PZT thin films. In *IEEE Proc. 8th Int. Symp. Appl. Ferroelectrics, Greenville, SC*, pages 244–247.

Wurfel, P. and Batra, I. P. (1974). Polarization reduction in thin ferroelectrics with semiconducting electrodes. *Ferroelectrics*, **7**, 261–263.

Xu, Yuhuan (1991). *Ferroelectric Materials and Their Applications.* Elsevier Science Publishers B.V., Amsterdam.

Xu, Yuhuan, Chen, Ching Ji, and Mackenzie, J. D. (1990). Self biased heterojunction effect of ferroelectric thin film on silicon. *Ferroelectrics*, **108**, 47–52.

Yan, Feng, Bao, Peng, Chan, Helen L. W., Choy, Chung-Loong, Chen, Xiaobing, Zhu, Jingsong, and Wang, Yening (2001). Effect of grain size on the fatigue properties of $Pb(Zr_{0.3}Ti_{0.7})O_3$ thin films prepared by metalorganic decomposition. *Ferroelectrics*, **252**, 209–216.

Yang, Jun-Kyu, Kim, Woo Sik, and Park, Hyung-Ho (2000a). The effect of excess Pb content on the crystallization and electrical properties in sol-gel derived $Pb(Zr_{0.4}Ti_{0.6})O_3$ thin films. *Thin Solid Films*, **377-378**, 739–744.

Yang, Jun-Kyu, Kim, Woo Sik, and Park, Hyung-Ho (2000b). Enhanced fatigue property through the control of interfacial layer in Pt/PZT/Pt structure. *Jpn. J. Appl. Phys.*, **39**, 7000–7002.

Yang, Jun-Kyu, Kim, Woo Sik, and Park, Hyung-Ho (2001). The investigation of Pb-sufficient buffer layer on the ferroelectric properties in Pt/PZT/Pt structure. *Ferroelectrics*, **260**, 267–272.

Yi, Insook, Kim, Daeik, Kim, Sukpil, Lee, Yongkyun, Kim, Changjung, and Chung, Ilsub (2001). Ferroelectric properties of sol-gel derived $Pb(Zr,Ti)O_3$ capacitors grown on IrO_2 electrode. *Integrated Ferroelectrics*, **37**, 135–44.

Yoo, In-K. and Desu, B. Seshu (1992a). Fatigue modeling of lead zirconate titanate thin films. *Mater. Sci. & Engin. B*, **13**, 319–322.

Yoo, In-K. and Desu, B. Seshu (1992b). Fatigue parameters of lead zirconate titanate thin films. *Mater. Res. Soc. Symp. Proc.*, **243**, 323–329.

Yoo, In-K. and Desu, S. B. (1992c). Mechanism of fatigue in ferroelectric thin films. *phys. stat. sol. (a)*, **133**, 565–573.

Yoo, In-K. and Desu, S. B. (1993). Fatigue and hysteresis modelling of ferroelectric materials. *J. Intell. Mater. Sys. Struct.*, **4**, 490–495.

Yoo, In-K., Desu, B. Seshu, and Xing, Jimmy (1993). Correlations among degradations in lead zirconate titanate thin film capacitors. *Mater. Res. Soc. Symp. Proc.*, **310**, 165–177.

Yoon, Ki Hyun, Shin, Hyun Cheol, Park, Jihoon, and Kang, Dong Heon (2002). Effect of textured $Pb(Zr_{1-x}Ti_x)O_3$ seed layer on fatigue properties of ferroelectric $Pb_{0.99}[(Zr_{0.6}Sn_{0.4})_{0.85}Ti_{0.15}]_{0.98}Nb_{0.02}O_3$ thin films. *J. Appl. Phys.*, **92**, 2108–2111.

Zammit-Mangion, L. J. and Saunders, G. A. (1984). Acoustic emission and domain wall dynamics in ferroelectric-ferroelastic gadolinium and terbium molybdate. *J. Phys. C: Solid State Phys.*, **17**, 2825–2830.

Zhang, N., Li, L., and Gui, Z. (2001a). Frequency dependence of ferroelectric fatigue in PLZT ceramics. *J. Eur. Ceram. Soc.*, **21**, 677–681.

Zhang, N., Li, L., and Gui, Z. (2001b). Improvement of electric fatigue properties in $Pb_{0.94}La_{0.04}(Zr_{0.70}Ti_{0.30})O_3$ ferroelectric capacitors due to $SrBi_2Nb_2O_9$ incorporation. *Mater. Res. Bull.*, **36**, 2553–2562.

Zhang, N., Li, L., and Gui, Z. (2001c). Influence of silver on electric properties in PLZT ferroelectrics ceramics. *J. Mater. Sci. Lett.*, **20**, 675–676.

Zhang, Ningxin, Li, Longtu, and Gui, Zhilun (2001d). Degradation of piezoelectric and dielectric relaxation properties due to electric fatigue in PLZT ferroelectric capacitors. *Mater. Lett.*, **48**, 39–45.

Zhang, Ningxin, Li, Longtu, Qi, Jianquan, and Gui, Zhilun (2001e). Frequency spectrum of ferroelectric fatigue in PLZT ceramics. *Ferroelectrics*, **259**, 109–114.

Zhang, Ningxin, Li, Longtu, and Gui, Zhilun (2001f). Improvement of electric fatigue in PLZT ferroelectric capacitors due to zirconia incorporation. *Mater. Chem. Phys.*, **72**, 5–10.

Zhang, Ningxin, Li, Longtu, and Gui, Zhilun (2001g). Improvement of electric fatigue properties in $Pb_{0.94}La_{0.04}(Zr_{0.70}Ti_{0.30})O_3$ ferroelectric capacitors due to $SrBi_2Nb_2O_9$ incorporation. *Mater. Res. Bul.*, **36**, 2553–2562.

Zhang, Y. and Jiang, Q. (1995). Twinning-induced internal stress in ferroelectric ceramics. *SPIE*, **2442**, 11–22.

Zhi, Jing, Chen, Ang, Zhi, Yu, Vilarinho, P. M., and Baptista, J. L. (1999). Incorporation of yttrium in barium titanate ceramics. *J. Am. Ceram. Soc.*, **82**, 1345–1348.

Zimmermann, A., Carter, W. C., and Fuller, Jr., E. R. (2001). Damage evolution during microcracking of brittle solids. *Acta Mater.*, **49**, 127–137.

Zorn, G., Wersing, W., and Göbel, H. (1985). Electrostrictive tensor components of PZT-ceramics measured by X-ray diffraction. *Jpn. J. Appl. Phys.*, **24**, 721–723.

Index

acceptor doping, 16
  unbalanced, 120
acoustic emission
  amplitudes, 90
    due to cracking, 91
    due to switching, 91
    table, 92
  barium titanate, 107
    needle domains, 109
  coincidence with Barkhausen pulses, 97, 106
  detection frequencies, 82
  domain annihilation, 109
  driving conditions, 93
  driving force level, 85
  electromagnetic noise, 85
  energy and amplitude estimates, 90
  energy distribution, 88, 89
    under compressive stress, 101
  field dependent patterns, 87
  gadolinium molybdate, 101, 104
    resonant frequencies, 104
    sound propagation direction, 103, 105
  measurement parameters, 82
  possible sources, 81
  principle, 81
  PZT, 85
  relaxation, 89
  setup transfer function, 82
  source frequencies, 82
  sources
    in ferroelectrics, 81, 94
    polycrystal, 84
  time dependent patterns, 86
  total energy, 90
  under additional uniaxial stress, 97
AFM, 182
agglomerate planes, 70
agglomerates, 63
  fields around, 151
  interaction with domain system, 70
agglomeration
  mechanism, 137
  model, 151
aging, 172
  and diffusion, 125
  at grain boundaries, 26
  by electronic carriers, 25
  defect-dipole reorientation, 23
  definition, 23
  Fe-doping, 24
  in barium titanate, 107
  point defect reordering, 27
  stabilized poling state, 26
amorphous grain boundary, 64
Anderson localization, 120
annealing of fatigue, 49
attempt frequency, 125

band edge, 120
band edges, 114
band structure, 114
  defect levels, 115
Barkhausen pulses
  ferroelectric, 84
  gadolinium molybdate, 104
  PZT, 98
bias field, 24, 47, 130
bismuth lanthanum titanate, 184
bismuth titanate, 184
BLT, 184
buffer layer
  antiferroelectric, 171
  MgO, 171
  PZT on Pt, 171

charge compensation
  by oxygen vacancies, 119
  by polarons, 119
  due to Y-doping, 39
charge relaxation, 130
chemical etching, 63
conducting dendritic paths, 68
conduction
  deep traps, 30
  grain boundary, 30
conductivity
  a.c., 119
crystallite orientation, 161

defect dipoles, 121
  alignment of, 121
defects
  short range interaction, 138
  strongly clamping, 21
  weakly clamping, 21
delamination, 78
dendritic conducting trees, 38
depolarizing field, 6
diffusion
  activation energy, 128
  anisotropy, 122
  bulk, 121
  directionality, 122
    with resp. to polarization direction, 123
  p-n-junction picture, 123
  silver, 65
diffusion barrier, 135
dipole fields, 118
dislocation loop
  association with agglomerates, 139
dislocation loops, partial, 139
domain
  clamping, 130
    degrees of, 131
  formation, 6
  freezing, 130
  nucleation inhibition, 34
domain density, 72
domain walls
  blocking of, 34
  clamping of 90°, 48
  curved, 176
domains
  blocked, 37
  fatigue affected, 54
  head to head, 36
  switching kinetics, 37
  tail to tail, 36, 141
  TEM, 176
  wedge shaped, 142, 153, 176
donor doping, 16
dopants, 8
  ionic radii of, 9

earthquake, 81
electrode
  near region, 171
  proximity, 63
  quality, 30
electrodes
  agglomeration of $V_O^{\bullet\bullet}$ at, 32, 38
  comb-shaped, 182
  majority carrier in, 38
  metal-, 27
    and fatigue, 28
    n- to p-type conversion by, 28
    tunneling, 29
  oxide, 168
    reducing atmosphere, 169
  oxide-, 30
    fatigue despite, 31
  rectifying p-n-junction, 31
  semiconductor-, 31
  silver, 65
  space charge underneath, 27
electrolyte electrodes, 118
electron mobility, 115
electron motion
  directionality, 129
electron orbitals, 20
electrostriction, 5, 46
embrittlement, 2
etch grooves, 63

fatigue
  -free SBT, 143
  -free systems, 39
  acoustic emissions due to, 96
  and aging, 172
  and Barkhausen pulses, 98
  anisotropy, 19, 51
  bulk, 32

definition, 1
due to point defect motion, 37
due to polarons, 159
induced phase change, 144
influence of crystal structure, 32
logarithmic, 36
macroscopic phenomenology, 18
mixed loading, 58
models, 36
Ising model, 172
new approaches, 172
phase transition due to, 70
rate effects, 34
saturation, 45
shift of Fermi-level, 135
single crystal orientation, 162
switching kinetics, 37, 38
temperature dependence, 33, 158
unipolar, 57
acoustic emission, 97
fatigue models
agglomerate, 112
aging, 112
crastallographic phase transition, 113
diffusion, 112
diffusion barrier, 113
electrodes, 113
field projections in polycrystal, 113
grain boundary, 112
ion drift, 113
iterative, 146
point defect, 111
structure, 111
fatigue setup
bipolar, 41
mixed electromechanical, 41
unipolar, 41
fields
modified, 130
frequency dependence
master curve, 181

grain boundary, 63, 136, 170
charge pile-up, 136
fields at, 7
Maxwell-Wagner relaxation, 136
strength, 67
grains
fatigue affected, 54

Hall effect, 115
hydrogen reduction, 169
hysteresis
fatigue-reduced, 43
ferroelectric, 4
loop deformations, 131
polarization, 43
slanted in thin films, 45
strain, 44
transverse strain, 52, 53

imprint, 169
imprint and aging, 26
imprint polarization, 130
incubation period, 45
internal field, 130
ion convection due to domain wall motion, 128
ion mobility, 127
ionic current, 147

Landau-Devonshire-theory, 4
lattice constant, 73
law of mass action, 15
layered perovskite ferroelectrics, 182
lead excess near electrode, 171
lead oxide
excess, 167
loss of, 167
volatility, 16, 67
lead vacancy, 14
lead zirconate titanate (PZT), 8
leakage current, 55
localized electron states, 118
logarithmic fatigue, 45

macrocracking, 78
measurement setup
polarization and strain, 42
mechanical fatigue, 178
microcracking, 35, 75, 179
anisotropic, 77
distribution, 78
TEM, 179
microdomains, 118
mixed loading, 2, 18
models
agglomeration, 38

morphotropic phase boundary, 8
multilayer actuator, 18
multiple ferroelectric layers, 171
multiple layers different composition, 171

offset
-field, 47, 130, 172
-polarization, 46, 47, 130, 172
equivalent dipoles, 48
oxygen
balance, 74
diffusion length, 75
partial pressure, 116
oxygen vacancy, 14
accumulation at electrodes, 135
charge state, 20
diffusion to grain boundary, 69
local order around, 117

Pb^{3+}-centers, 120
PIC151, 10
piezoelectric coefficient, 5
piezoelectricity, 5
pinned domains, 48
platinum oxide, 169
point defect
equilibria, 8, 14
in unit cell, 20
point defects, 14
at grain boundary, 17
immobile pairs, 117
in $BaTiO_3$, $SrTiO_3$, 15
in fatigue, 20
quenched, 16
polarization
loss, 43
non-switchable, 37
stretched exponential loss, 181
polaron, 119
Preisach model, 172
PZT
(2% La) hysteresis loops, 12
(2% La) material parameters, 13
hysteresis, 11
microstructure, 10
phase diagram, 8
rhombohedral, 12
tetragonal, 12

rejuvenation, 165
high field-, 49
relaxation, 55
relaxation times, 164
relevant authors, 15, 18, 31

sample coloring, 55
sample self-heating, 35
SBT, 182
aging, 182
Schottky barrier, 27
Schottky defects, 14
screening, 6, 130
length, 28
time, 150
secondary bulk phases, 168
silver
incorporation, 67
migration, 68
oxidation, 67
space charge
$V_O^{\bullet\bullet}$, 27
strain, 5
asymmetry, 19, 46
loss, 43
mismatch, 78
stress
internal, due to transverse volume change, 54
stretched exponential recovery, 22
strontium bismuth tantalate, 182
switchable polarization
in polycrystals, 131
power law loss of, 148
reduction of, 18
switching
Gibb's free energy asymmetry, 132
switching mechanism
and acoustic emissions, 101

TEM, 63, 176
tetragonality, enhanced, 145
texture due to fatigue, 33
thermally stimulated short circuit current, 133
Ti^{3+}-centers, 120
trap states, 22
triple points, 64

uniaxial ferroelectrics, 185
unit cell volume, 73
UV illumination, 22
+ applied field, 22
+ heat treatment, 22
domain clamping, 132

vacancy clusters, planar, 139
vacancy pair $\{V_O^{\bullet\bullet}\text{-}V_{Pb}''\}$, 167, 171

Springer Series in
MATERIALS SCIENCE

10 **Computer Simulation of Ion-Solid Interactions**
By W. Eckstein

11 **Mechanisms of High Temperature Superconductivity**
Editors: H. Kamimura and A. Oshiyama

12 **Dislocation Dynamics and Plasticity**
By T. Suzuki, S. Takeuchi, and H. Yoshinaga

13 **Semiconductor Silicon**
Materials Science and Technology
Editors: G. Harbeke and M. J. Schulz

14 **Graphite Intercalation Compounds I**
Structure and Dynamics
Editors: H. Zabel and S. A. Solin

15 **Crystal Chemistry of High-T_c Superconducting Copper Oxides**
By B. Raveau, C. Michel, M. Hervieu, and D. Groult

16 **Hydrogen in Semiconductors**
By S. J. Pearton, M. Stavola, and J. W. Corbett

17 **Ordering at Surfaces and Interfaces**
Editors: A. Yoshimori, T. Shinjo, and H. Watanabe

18 **Graphite Intercalation Compounds II**
Editors: S. A. Solin and H. Zabel

19 **Laser-Assisted Microtechnology**
By S. M. Metev and V. P. Veiko
2nd Edition

20 **Microcluster Physics**
By S. Sugano and H. Koizumi
2nd Edition

21 **The Metal-Hydrogen System**
By Y. Fukai

22 **Ion Implantation in Diamond, Graphite and Related Materials**
By M. S. Dresselhaus and R. Kalish

23 **The Real Structure of High-T_c Superconductors**
Editor: V. Sh. Shekhtman

24 **Metal Impurities in Silicon-Device Fabrication**
By K. Graff 2nd Edition

25 **Optical Properties of Metal Clusters**
By U. Kreibig and M. Vollmer

26 **Gas Source Molecular Beam Epitaxy**
Growth and Properties of Phosphorus Containing III–V Heterostructures
By M. B. Panish and H. Temkin

27 **Physics of New Materials**
Editor: F. E. Fujita 2nd Edition

28 **Laser Ablation**
Principles and Applications
Editor: J. C. Miller

29 **Elements of Rapid Solidification**
Fundamentals and Applications
Editor: M. A. Otooni

30 **Process Technology for Semiconductor Lasers**
Crystal Growth and Microprocesses
By K. Iga and S. Kinoshita

31 **Nanostructures and Quantum Effects**
By H. Sakaki and H. Noge

32 **Nitride Semiconductors and Devices**
By H. Morkoç

33 **Supercarbon**
Synthesis, Properties and Applications
Editors: S. Yoshimura and R. P. H. Chang

34 **Computational Materials Design**
Editor: T. Saito

35 **Macromolecular Science and Engineering**
New Aspects
Editor: Y. Tanabe

36 **Ceramics**
Mechanical Properties, Failure Behaviour, Materials Selection
By D. Munz and T. Fett

37 **Technology and Applications of Amorphous Silicon**
Editor: R. A. Street

38 **Fullerene Polymers and Fullerene Polymer Composites**
Editors: P. C. Eklund and A. M. Rao

Springer Series in
MATERIALS SCIENCE

Editors: R. Hull R. M. Osgood, Jr. J. Parisi H. Warlimont

39 **Semiconducting Silicides**
Editor: V. E. Borisenko

40 **Reference Materials in Analytical Chemistry**
A Guide for Selection and Use
Editor: A. Zschunke

41 **Organic Electronic Materials**
Conjugated Polymers and Low Molecular Weight Organic Solids
Editors: R. Farchioni and G. Grosso

42 **Raman Scattering in Materials Science**
Editors: W. H. Weber and R. Merlin

43 **The Atomistic Nature of Crystal Growth**
By B. Mutaftschiev

44 **Thermodynamic Basis of Crystal Growth**
P–T–X Phase Equilibrium and Non-Stoichiometry
By J. Greenberg

45 **Thermoelectrics**
Basic Principles and New Materials Developments
By G. S. Nolas, J. Sharp, and H. J. Goldsmid

46 **Fundamental Aspects of Silicon Oxidation**
Editor: Y. J. Chabal

47 **Disorder and Order in Strongly Nonstoichiometric Compounds**
Transition Metal Carbides, Nitrides and Oxides
By A. I. Gusev, A. A. Rempel, and A. J. Magerl

48 **The Glass Transition**
Relaxation Dynamics in Liquids and Disordered Materials
By E. Donth

49 **Alkali Halides**
A Handbook of Physical Properties
By D. B. Sirdeshmukh, L. Sirdeshmukh, and K. G. Subhadra

50 **High-Resolution Imaging and Spectrometry of Materials**
Editors: F. Ernst and M. Rühle

51 **Point Defects in Semiconductors and Insulators**
Determination of Atomic and Electronic Structure from Paramagnetic Hyperfine Interactions
By J.-M. Spaeth and H. Overhof

52 **Polymer Films with Embedded Metal Nanoparticles**
By A. Heilmann

53 **Nanocrystalline Ceramics**
Synthesis and Structure
By M. Winterer

54 **Electronic Structure and Magnetism of Complex Materials**
Editors: D.J. Singh and D. A. Papaconstantopoulos

55 **Quasicrystals**
An Introduction to Structure, Physical Properties and Applications
Editors: J.-B. Suck, M. Schreiber, and P. Häussler

56 **SiO_2 in Si Microdevices**
By M. Itsumi

57 **Radiation Effects in Advanced Semiconductor Materials and Devices**
By C. Claeys and E. Simoen

58 **Functional Thin Films and Functional Materials**
New Concepts and Technologies
Editor: D. Shi

59 **Dielectric Properties of Porous Media**
By S.O. Gladkov

60 **Organic Photovoltaics**
Concepts and Realization
Editors: C. Brabec, V. Dyakonov, J. Parisi and N. Sariciftci

Printing: Saladruck Berlin
Binding: Lüderitz&Bauer, Berlin

CPSIA information can be obtained
at www.ICGtesting.com
Printed in the USA
LVHW051043030520
654914LV00002B/374

9 783540 402350